T0210957

Communications in Computer and Information Science   **941**

Commenced Publication in 2007
Founding and Former Series Editors:
Phoebe Chen, Alfredo Cuzzocrea, Xiaoyong Du, Orhun Kara, Ting Liu,
Dominik Ślęzak, and Xiaokang Yang

More information about this series at http://www.springer.com/series/7899

Leman Akoglu · Emilio Ferrara
Mallayya Deivamani · Ricardo Baeza-Yates
Palanisamy Yogesh (Eds.)

Advances in Data Science

Third International Conference
on Intelligent Information Technologies, ICIIT 2018
Chennai, India, December 11–14, 2018
Proceedings

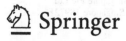
Springer

Editors
Leman Akoglu
Carnegie Mellon University
Pittsburgh, PA, USA

Ricardo Baeza-Yates
Northeastern University at Silicon Valley
San Jose, CA, USA

Emilio Ferrara
University of Southern California
Marina Del Rey, CA, USA

Palanisamy Yogesh
Anna University
Chennai, India

Mallayya Deivamani
CEG
Anna University
Chennai, India

ISSN 1865-0929 ISSN 1865-0937 (electronic)
Communications in Computer and Information Science
ISBN 978-981-13-3581-5 ISBN 978-981-13-3582-2 (eBook)
https://doi.org/10.1007/978-981-13-3582-2

Library of Congress Control Number: 2018962935

This Springer imprint is published by the registered company Springer Nature Singapore Pte Ltd.
The registered company address is: 152 Beach Road, #21-01/04 Gateway East, Singapore 189721, Singapore

General Chairs' Preface

On behalf of the Organizing Committee, we are pleased to welcome you to the proceedings of the Third International Conference on Intelligent Information Technologies (ICIIT 2018) organized by the Department of Information Science and Technology, College of Engineering Guindy (CEG), Anna University Chennai, India. ICIIT successfully brought together researchers and developers, with the purpose of identifying challenging problems in recent technologies.

We were delighted to present four outstanding keynote speakers: Dr. C. Mohan IBM Almaden Research Center, USA; Prof. Jure Leskovec from Stanford University, USA; Prof. Raj Reddy from Carnegie Mellon University, USA; and Dr. Rajeev Rastogi from Amazon, India.

ICIIT 2018 captivated with a signature event – "Industry Day" – to share practices among academia and industry. The industry keynote speakers were: Dr. Lipika Dey from Tata Consultancy Services, India; Amruta Joshi from Google, India; Hari Vasudev from Walmart Labs, India; and Ravi Vijayaraghavan from Flipkart, India.

We are grateful to the many authors who submitted their work to the ICIIT technical program. The Program Committee was led by Leman Akoglu and Emilio Ferrara. A report on the paper selection process appears in the PC Chairs' Preface.

We also thank the other chairs in the organization team: Prof. Saswati Mukherjee for acting as a convener of the conference; Dr. Mallayya Deivamani for publicizing the event to attract submissions and for managing the website, handling the proceedings process, and the local arrangements, thus ensuring the conference ran smoothly; Prof. Swamynathan Sankara Narayanan for acting as finance chair; Dr. Muthusamy Chelliah, Flipkart, India for acting as panel chair; and Dr. Shalini Urs, MYRA School of Business, Mysuru, India for acting as tutorial chair.

We are grateful to the sponsors of the conference, Indian Space Research Organization (ISRO), and Flipkart India, for their generous sponsorship and support. We would also like to express our gratitude to the College of Engineering (CEG), Anna University Chennai for hosting and organizing this conference. Last but not least, our sincere thanks go to all the local team members and volunteer helpers for their hard work to make the event possible. We hope you enjoy the proceedings of ICIIT 2018.

Ricardo Baeza-Yates
Palanisamy Yogesh

PC Chairs' Preface

On behalf of the Program Committee, it is our pleasure to present to you the proceedings of the International Conference on Intelligent Information Technologies (ICIIT 2018) held during December 11–13, 2018, at the College of Engineering Guindy, Anna University Chennai, India. ICIIT 2018 acted as a forum for the researchers, scientists, academics and industrialists to present their latest research results and research perspectives on the conference theme, "Data Science and Analytics."

The conference received 74 submissions from all over the world. After a rigorous peer-review process involving 235 reviews in total, 15 full-length articles were accepted for oral presentation and for inclusion in the CCIS proceedings. This corresponds to an acceptance rate of 20% and is intended for maintaining the high standards of the conference proceedings. The papers included in this CCIS volume cover a wide range of topics in data science foundations, data management and processing technologies, and data analytics and its applications.

The pre-conference tutorials conducted on December 10, 2018, covered the thrust areas of data science and analytics. The technical program started on December 11, 2018, and continued for two days. Non-overlapping oral and poster sessions ensured that all attendees had the opportunity to interact personally with presenters. The conference featured distinguished keynote speakers, Dr. Mohan Chandrasekaran of IBM, USA, Prof. Jure Leskovec of Stanford University, USA, Prof. Raj Reddy of Carnegie Mellon University, USA, and Dr. Rajeev Rastogi of Amazon, India.

We take this opportunity to thank the authors of all submitted papers for their hard work, adherence to the deadlines, and patience with the review process. The quality of a refereed volume depends mainly on the expertise and dedication of the reviewers. We are thankful to the reviewers for their timely effort and help to make this conference successful. We, thank Prof. Ricardo Bazea-Yates of NTENT and Northeastern University at SV, USA, and Prof. Palanisamy Yogesh of Anna University, Chennai, for providing valuable guidelines and inspiration to overcome various difficulties in the process of organizing this conference as general co-chairs. We would like to thank the track chairs – Dr. Suren Byna of Lawrence Berkeley National Laboratory, USA; Dr. Amruta Joshi of Google, India; Dr. Eleanor Loh of Deliveroo, UK; Dr. Mallayya Deivamani of College of Engineering Guindy (CEG), India; Dr. Moumita Sinha of Adobe, USA; and Dr. Ravi Vijayaraghavan of Flipkart, India – for their effort toward the review process of ICIIT 2018. We thank Prof. Saswati Mukherjee and Prof. Swamynathan S. for their endless effort in all aspects as conference convener and finance chair, respectively. For the publishing process at Springer, we would like to thank, Leonie Kunz, Yeshmeena Bisht, Suvira Srivastav, and Nidhi Chandhoke for their constant help and cooperation.

Our sincere and heartfelt thanks to Prof. M. K. Surappa, vice-chancellor of Anna University, Chennai, Prof. J. Kumar, Registrar of Anna University, Chennai, and Prof.

Geetha T V, Dean, College of Engineering (CEG), Anna University, for their support toward ICIIT 2018 and providing the infrastructure at CEG to organize the conference. We are indebted to the faculty, staff, and students of the Department of Information Science and Technology for their tireless efforts that made ICIIT 2018 at CEG possible. We would also like to thank the sponsors Indian Space Research Organization (ISRO) and Flipkart for their support. We would also like to thank the participants of this conference, who have considered the conference above all hardships. In addition, we would like to express our appreciation and thanks to all the people whose efforts made this conference a grand success.

December 2018

Leman Akoglu
Emilio Ferrara

Organization

ICIIT 2018 was organized by the Department of Information Science and Technology, College of Engineering Guindy, Anna University, Chennai, India.

Chief Patron

Surappa M. K.   Anna University, India
   (Vice-chancellor)

Patrons

Kumar J. (Registrar)   Anna University, India
Geetha T. V. (Dean)   CEG, Anna University, India

General Co-chairs

Ricardo Baeza-Yates   NTENT and Northeastern University, USA
Palanisamy Yogesh   CEG, Anna University, India

Program Co-chairs

Leman Akoglu   Carnegie Mellon University, USA
Emilio Ferrara   University of Southern California, USA

Track Chairs

Suren Byna   Lawrence Berkeley National Laboratory, USA
Amruta Joshi   Google, India
Eleanor Loh   Deliveroo, UK
Mallayya Deivamani   CEG, Anna University, India
Moumita Sinha   Adobe, USA
Ravi Vijayaraghavan   Flipkart, India

Convener

Saswati Mukherjee   CEG, Anna University, India

Local Arrangements Chair and Proceedings Chair

Mallayya Deivamani   CEG, Anna University, India

Finance Chair

Swamynathan Sankara   CEG, Anna University, India
  Narayanan

Panel Chair

Muthusamy Chelliah   Flipkart, India

Tutorial Chair

Shalini Urs   MYRA School of Business, Mysuru, India

Organizing Committee

Ranjani Parthsarathi	Bama Srinivasan
Uma G. V.	Sairamesh L.
Indhumathi J.	Pandiyaraju V.
Sridhar S.	Vijaykumar T. J.
Vani K.	Narashiman D.
Geetha Ramani R.	Prabhavathy P.
Mala T.	Shunmuga Perumal P.
Sendhil Kumar S.	Ezhilarasi V.
Kulothungan K.	Tina Esther Trueman
Vijayalakshmi M.	Kanimozhi S.
Thangaraj N.	Sindhu T.
Indra Gandhi K.	Senthilnayaki B.
Uma E.	Jasmine R. L.
Abirami S.	Riasudheen H.
Geetha P.	Yuvaraj B. R.
Vidhya K.	Mohana Bhindu K.
Selvi Ravindran	Mahalakshmi G.
Muthuraj R.	

Technical Review Board

Abdullah Tansel   The City University of New York, USA
Akhilesh Bajaj   The University of Tulsa, Oklahoma
Alberto Cano   Virginia Commonwealth University, USA
Alfredo Cuzzocrea   University of Trieste, Italy
Andrea Clematis   IMATI-CNR, Italy
Antonis Sidiropoulos   Aristotle University of Thessaloniki, Greece
Azad Naik   Microsoft Research, USA
Bharath Balasubramanian   AT & T Labs, USA

Bolong Zheng	Aalborg University, Denmark
Chitra Babu	SSN College of Engineering, India
David Lillis	University College Dublin, Ireland
Dharavath Ramesh	Indian Institute of Technology – Dhanbad, India
Dhiraj Sangwan	CSIR-CEERI, Pilani, India
Ehsan Ullah	Qatar Computing Research Institute, Qatar
Eleni Mangina	University College Dublin, Ireland
Felix Gessert	University of Hamburg, Germany
Feng Yan	University of Nevada, Reno, USA
G. C. Nandi	IIIT – Allahabad, India
Grigori Sidorov	National Polytechnic Institute, IPN, Mexico
Guilherme Desouza	University of Missouri, USA
Gunter Saake	University of Magdeburg, Germany
Han Fang	Facebook, USA
Hasan Kurban	Indiana University, USA
Hiba Arnaout	Saarland Informatics Campus, Germany
Houjun Tang	Lawrence Berkeley National Lab (LBNL), USA
Hu Chun	Google Inc., USA
Jang Hyun Kim	Sungkyunkwan University, South Korea
Jay Lofstead	Sandia National Laboratories, California, USA
Jesús Camacho-Rodríguez	Hortonworks Inc., USA
Jian Wu	The Pennsylvania State University, USA
Jiawen Yao	University of Texas at Arlington, USA
Jingchao Ni	The Pennsylvania State University, USA
Ka-Chun Wong	City University of Hong Kong, SAR China
Kalidas Yeturu	Indian Institute of Technology – Tirupati, India
Kanchana R.	SSN College of Engineering, India
Krishnaprasad Thirunarayan	Kno.e.sis Center, Wright State University, USA
Laurent Anne	University of Montpellier, LIRMM, CNRS, France
Laurent D'Orazio	University of Rennes, France
Li-Shiang Tsay	North Carolina A & T State University, USA
Liting Hu	Florida International University, USA
Manar Mohammed	Miami University, USA
Manas Gaur	Kno.e.sis Center, Wright State University, USA
Manoj Thuasidas	Singapore Management University, Singapore
Marijn Ten Thij	VU University Amsterdam, The Netherlands
Mehmet Dalkilic	Indiana University, USA
Michele Melchiori	University of Brescia, Italy
Mike Jackson	Birmingham City University, USA
Mingjie Tang	Hortonworks, USA
Mirco Schoenfeld	HFP/TU Munich, Germany
Mohammad Haque	The University of Newcastle, Australia
Murat Ünalır	Ege University, Turkey
Pradeep Kumar	IIM – Lucknow, India

Additional Reviewers

Bing Xie	Oak Ridge National Lab, USA
Dongkuan Xu	The Pennsylvania State University, USA
Felix Enigo	SSN College of Engineering, India
Kumar Vinayak	IIT Hyderabad, India
Özgü Can	Ege University, Turkey
Udit Arora	IIITD, India
Weiqing Yang	Hortonworks, USA
Yanbo Liang	Hortonworks, USA
Yongyang Yu	Purdue University, USA
Yu Hao	UNSW, Australia

Keynotes

Blockchains Untangled: Public, Private, Smart Contracts, Applications, Issues

C. Mohan

IBM Almaden Research Center, USA

Abstract. The concept of a distributed ledger was invented as the underlying technology of the public or permission less Bitcoin cryptocurrency network. But the adoption and further adaptation of it for use in the private or permissioned environments is what I consider to be of practical consequence and hence only such private blockchain systems will be the focus of this talk.

Computer companies like IBM, Intel, Oracle, Baidu and Microsoft, and many key players in different vertical industry segments have recognized the applicability of blockchains in environments other than cryptocurrencies. IBM did some pioneering work by architecting and implementing Fabric, and then open sourcing it. Now Fabric is being enhanced via the Hyperledger Consortium as part of The Linux Foundation. There is a great deal of momentum behind Hyperledger Fabric throughout the world. Other private blockchain efforts include Enterprise Ethereum, Hyperledger Sawtooth and R3 Corda.

While currently there is no standard in the private blockchain space, all the ongoing efforts involve some combination of persistence, transaction, encryption, virtualization, consensus and other distributed systems technologies. Some of the application areas in which blockchain systems have been leveraged are: global trade digitization, derivatives processing, e-governance, Know Your Customer (KYC), healthcare, food safety, supply chain management and provenance management.

In this talk, I will describe some use-case scenarios, especially those in production deployment. I will also survey the landscape of private blockchain systems with respect to their architectures in general and their approaches to some specific technical areas. I will also discuss some of the opportunities that exist and the challenges that need to be addressed. Since most of the blockchain efforts are still in a nascent state, the time is right for mainstream database and distributed systems researchers and practitioners to get more deeply involved to focus on the numerous open problems. Extensive blockchain related collateral can be found at http://bit.ly/CMbcDB.

Graph Representation Learning

Jure Leskovec

Stanford University, USA

Abstract. Machine learning on graphs is an important and ubiquitous task with applications ranging from drug design to friendship recommendation in social networks. The primary challenge in this domain is finding a way to represent, or encode, graph structure so that it can be easily exploited by machine learning models. However, traditionally machine learning approaches relied on user-defined heuristics to extract features encoding structural information about a graph. In this talk I will discuss methods that automatically learn to encode graph structure into low-dimensional embeddings, using techniques based on deep learning and nonlinear dimensionality reduction. I will provide a conceptual review of key advancements in this area of representation learning on graphs, including random-walk based algorithms, and graph convolutional networks.

AI: Background, History and Future Opportunities

Raj Reddy

Carnegie Mellon University, USA

Abstract. This talk will provide Background and History of AI in an attempt to clarify the sources of misinformation about AI in the media recently. Many of these predictions are based on flawed reasoning and incorrect extrapolations and will not happen. Robots will not take over the world. In this talk, we will review tools, techniques and advances in AI over the past half century and explore what might be next. We will discuss how these Intelligent Agents will lead to Knowledge as a Service Industry (KaaS) and create a Market Place for Apps that provide KaaS.

Machine Learning @ Amazon

Rajeev Rastogi

Amazon, India

Abstract. In this talk, I will first provide an overview of key problem areas where we are applying Machine Learning (ML) techniques within Amazon such as product demand forecasting, product search, and information extraction from reviews, and associated technical challenges. I will then talk about three specific applications where we use a variety of methods to learn semantically rich representations of data: question answering where we use deep learning techniques, product size recommendations where we use probabilistic models, and fake reviews detection where we use tensor factorization algorithms.

Industrial Invited Talks

Designing Automated Decision-Making Systems – Today, Tomorrow and the Day After

Lipika Dey

Tata Consultancy Services, India

Abstract. Recent advances in AI technologies backed by the success of Siri, Facebook, Google and Alexa have raised the expectations of Enterprises to automate several routine tasks that involved dealing with unstructured information from Video, Speech and Text data and heretofore were exclusively dealt with by human agents. While there is indeed enough scope to employ AI technologies in conjunction with Data Analytics, designing such systems need to address many more issues other than the accuracies and performances of the underlying analytics algorithms. Other than addressing increasing concerns about privacy and security for such systems, ensuring learnability, trustability and explainability are some of the crucial questions that are increasingly coming up. This talk will discuss about these concerns and some attempts that are being taken to ensure order the madness.

Data Sciences in the Cloud

Amruta Joshi

Google, India

Abstract. With growing awareness of the importance of Data Sciences, the field is getting more and more sophisticated. We now regularly process larger amounts of data and use more complex algorithms. This new landscape comes with its own set of challenges: it requires large-scale infrastructure, robust multi-level security, and more sophisticated multi-feature algorithms to process the data. Cloud provides powerful tools to tackle all of these challenges. In this talk we will see how Data Science and Cloud work together with each other to help you manage your big data and convert it into meaningful insights at large scale.

Using Big Data to Change the Way the World Shops

Hari Vasudev

Walmart Labs, India

Abstract. With the growing convergence of Big Data, ML, AI and Cloud, there is a huge transformation underway in Retail today that is impacting Customers, Merchants, global Supply Chains and Employees. Walmart's vision for Big Data is to deliver data-driven experiences that help customers save money (and time) so they can live better. In this talk, we'll explore how Big Data algorithms and sciences is being used at the world's largest retailer to fundamentally change the way the world shops and deliver unmatched omni-channel shopping experiences.

Analytics and Decision Sciences for Ecommerce - An Overview and Use Cases on Pricing and Selection

Ravi Vijayaraghavan

Flipkart, India

Abstract. The Analytics and Decision Sciences organisation at Flipkart has the charter of leveraging science to enable robust data-driven decision-making. Key focus areas of this organisation are business growth and continuously improving customer experience. My talk will present the overall landscape of the Analytics organisation and the areas we cover.

Two key expectations of any consumer from an ecommerce platform are –

1. The availability of a really wide assortment/selection of products and
2. A price that is true value for money

Following the overview, I will present use cases related to applications of statistics, machine learning and optimisation in addressing these two expectations of our consumers.

Adoption of Analytics in Engineering

Srinath Jangam

L & T Construction, Inc., India

Abstract. Artificial intelligence is poised to unleash the next wave of digital disruption, and companies should prepare for it now. There are real-life benefits for a few early adopting firms, making it more Important than ever all sectors to accelerate and adoption to their digital/IOT transformations. Five AI technology systems to focous is on robotics and autonomous vehicles, computer vision, language, virtual agents, and machine learning, which includes deep learning and underpins many recent advances in the other AI technologies. Early evidence suggests that AI can deliver real value to serious adopters and can be a powerful force for disruption. Early AI adopters that combine strong digital/IOT capability with proactive strategies have higher profit margins and expect the performance gap with other firms to widen in the future. AI promises benefits, but also poses urgent challenges that cut across firms, developers, government, and workers. The workforce needs to be reskilled to exploit AI rather than compete with it.

Tutorials

Big Data or Right Data?
Opportunities and Challenges

Ricardo Baeza-Yates

NTENT and Northeastern University at SV, USA

Abstract. Big data nowadays is a fashionable topic, independently of what people mean when they use this term. But being big is just a matter of volume, although there is no clear agreement in the size threshold. On the other hand, it is easy to capture large amounts of data using a brute force approach. So, the real goal should not be big data but to ask ourselves, for a given problem, what is the right data and how much of it is needed. For some problems, this would imply big data, but for most of the problems much less data will and is needed. Hence, in this presentation, we cover the opportunities and the challenges behind big data. Regarding the challenges, we explore the trade-offs involved with the main problems that arise with big data: scalability, redundancy, bias, the bubble filter and privacy.

Deep Learning Models
for Image Processing Tasks

C. ChandraSekhar

Indian Institute of Technology Madras, India

Abstract. The shallow learning models based on conventional machine learning techniques for pattern classification such as Gaussian mixture models, multilayer feedforward neural networks and support vector machines use the hand-picked features as input to the models. Recently, several deep learning models have been explored for learning a suitable representation from the image data and then using the learnt representation for performing the image pattern analysis tasks such as image classification, annotation and captioning. In this talk, we present the deep learning models such as Stacked autoencoder, Deep convolutional neural network and Stacked restricted Boltzmann machine for learning a suitable representation from the image data. Then, we present the deep learning models-based approaches to image classification, image annotation and image captioning.

Signal Processing guided Machine Learning

Hema A. Murthy

Indian Institute of Technology Madras, India

Abstract. Machine learning has become ubiquitous today. Big data analytics has become the buzzword. Build, train, and deploy/transfer is the paradigm that has become the "mantra" today. The more the amount of data available, the more robust the systems are at prediction. The major problem with machine learning is the problem of getting huge amount data that has been curated. In the context of speech technologies, in a country like India with a large linguistic diversity, getting data that is accurate for training is difficult. Another issue, is that of simultaneous collection of data in multiple languages. Is there a way to reduce the amount of data required for training a machine learning system? In this tutorial, we show how signal processing can be used to guide machine learning algorithms. In particular we study problems in speech synthesis, recognition, Indian music analysis, and computational brain research, where efforts are made to first process the signal before subjecting it to machine learning. Signal processing yields accurate results in the particular, while it may lead to a large number of insertions, deletions, and substitutions. Using machine learning, and signal processing in tandem we show that the amount of data required for training systems can be reduced significantly.

Visual Analytics: "Bringing Data to Life"

Jaya Sreevalsan Nair

International Institute of Information Technology Bangalore, India

Abstract. John Tukey, the mathematician, said the following. once upon a time about analytics: "This is my favorite part about analytics: Taking boring flat data and bringing it to life through visualization." It remains true to a great extent even today, in the time of big data. The objective of this tutorial is to impress upon the audience the need for visualization as an essential part of larger data science workflows. Visualization in itself has evolved from being summaries to facilitating complex exploratory analysis of data. This tutorial will demonstrate techniques of how data can be formatted to make the best use of some of the time-tested visualization techniques, and how visualizations enable in the overall data analysis.

Biological/Genomic Data Science: Moving Beyond Correlation to Causation

Manikandan Narayanan

Indian Institute of Technology Madras, India

Abstract. Discovering causal relations in a complex system is a fundamental pursuit in many sciences and disciplines. When controlled intervention experiments to determine cause-and-effect is not feasible or ethical, causal inference is surprisingly possible from observational data alone - its theory (models/ assumptions/language) and practice (concrete applications in biology/medicine) is the focus of this tutorial. You will find this tutorial appealing if you find causal inference from observational data intriguing (e.g., how can one break the symmetry of an observed correlation between two variables to determine the causal direction, or sever the links to not only known but also unknown confounding factors?) and valuable (in terms of its broad applications, including bioinformatics applications ranging from identifying causal risk factors of human diseases to gene regulatory networks underlying living cells).

We will start with causal discovery between two variables using the so-called mediation-based and Mendelian Randomization (MR) approaches that are analogous to Randomized Controlled Trials popularized by Ronald Fisher, and then move onto multivariate causal discovery using the framework of Bayesian networks and do-calculus pioneered by Judea Pearl. We intend to cover modern developments and data resources that aid causal discovery from biomedical/genomic data (for instance, one recent resource pools 11 billion correlations between genetic variants and health/disease-related outcomes from genome-wide association studies, which are waiting to be mined for new causal factors for human health and disease).

All relevant biology and causality concepts will be introduced. A basic knowledge of probability/statistics is assumed.

Social Network Analysis:
Making the Invisible Visible

Shalini R. Urs

MYRA School of Business, Mysore, India

Abstract. Over the past decade, there has been a growing public fascination with the complex "connectedness" of modern society especially since the emergence of Social Networking sites. Whether the rapid spread of news or the tipping point of social/political movements gathering momentum or the cascading of epidemics and financial crises around the world with alacrity and intensity, it is attributed to the connectedness of today's society. Many scientific disciplines have come together and evolved into a new discipline of network science focused on understanding these complex connected systems operate. Social Network Analysis (SNA) has emerged as an approach and a tool to uncover and understand the hidden side of connections. This tutorial will introduce the basic concepts of a network, their attributes and their measures such as Centrality, Components, Cohesion, Geodesic, Density and Degree, Cores, Cliques and others. We will also introduce Graph Theory that underpins network science and uses graph theory as a primary tool in the broader examination of networks. With the help of examples from across different domains, we will help participants understand the dynamics of social networks and how this understanding can be used from uncovering terrorist networks to "The Network of Global Corporate Control." This tutorial will introduce the participants to some of the essential software tools such as Gephi, Pajek, NodeXL, Cytoscape and NetworkX. Participants will be shown how to install, import data and analyze the network with the help of examples. A comparison of these five software tools concerning features and performance will be presented.

Contents

Data Science Foundations

GCRITICPA: A CRITIC and Grey Relational Analysis Based Service Ranking Approach for Cloud Service Selection

Gireesha Obulaporam[1], Nivethitha Somu[2],
Gauthama Raman ManiIyer Ramani[3], Akshya Kaveri Boopathy[1],
and Shankar Sriram Vathula Sankaran[1(✉)]

[1] Centre for Information Super Highway (CISH), School of Computing,
SASTRA Deemed to be University, Thanjavur 613401, Tamil Nadu, India
sriram@it.sastra.edu
[2] Smart Energy Informatics Laboratory (SEIL), Indian Institute of Technology-
Bombay, Mumbai 400076, Maharashtra, India
[3] iTrust, Centre for Research in Cyber Security,
Singapore University of Technology and Design (SUTD),
Singapore City, Singapore

Abstract. The dynamic nature of users' functional & non-functional requirements and the ever-increasing number of cloud service providers with similar
functionalities poses a significant challenge on the identification of trustworthy
cloud services. Existing service selection approaches lacks in trust-based evaluation in terms of compliance between the service provision and the user
requirements mentioned in the Service Level Agreements (SLAs), which results
in inconsistent ranking of Cloud Service Providers (CSPs) due to the rank
reversal phenomenon. As a solution to the rank reversal problem, this paper
presents Grey CRiteria Importance Through Intercriteria Correlation Parallel
Analysis (GCRITICPA) based service ranking approach that employs CRITICA
to determine weights of the criteria Grey Relational Analysis (GRA) to rank the
trustworthy cloud service providers. A case study using real world Cloud Armor
trust feedback dataset demonstrates the efficiency of the proposed approach in
terms of trustworthiness and rank preservation.

Keywords: Cloud Service Ranking · CRITICA · GRA · SLA
Trustworthy cloud service providers · Rank reversal

1 Introduction

The rapid development of novel and efficient computing methods attracted the practitioners and various enterprises towards Internet-based computing such as Internet of
Things (IoT), cloud computing, and Cloud of Things (CoT). Cloud computing, a
distributed computing model facilitates the access to highly reliable, scalable, and
flexible on demand computing resources over the internet in a 'Pay-As-You-Use' basis
[1]. The immense popularity of cloud computing and the upsurge of numerous cloud
service providers offering functionally-equivalent cloud services at different levels of

© Springer Nature Singapore Pte Ltd. 2019
L. Akoglu et al. (Eds.): ICIIT 2018, CCIS 941, pp. 3–16, 2019.
https://doi.org/10.1007/978-981-13-3582-2_1

abstraction, performance and pricing policies makes it difficult for the users to identify the appropriate and suitable cloud service providers. The diverse nature of cloud services and absence of service-publication benchmarks makes it difficult for the cloud users and decision makers to selection a trustworthy and user requirement compliant cloud service. Trust based cloud service selection models have gained its attraction among the researchers for the design of a robust and reliable cloud service selection model [2] since the trustworthiness of the cloud service providers is a measure of their quality based on several QoS attributes. According to Somu et al., "Trustworthiness, an important quality metric of a service provider, depends on various trust measure parameters (TMPs) such as availability, accountability and cost which can be measured using objective and subjective assessment techniques" [3–5].

A Cloud Service Selection (CSS) problem is viewed as a Multi Criteria Decision Making (MCDM) problem since it involves intrinsic relationship among the QoS attributes. Over the past few decades, several MCDM approaches were utilized to rank the cloud service providers (Table 1) however, they fail to address the rank reversal problem [6, 7]. The rank reversal problem was first identified by Belton and Gear in 1983 [8] and is defined as the general phenomenon which occurs during the addition/deletion of new alternatives that leads to noticeable changes in the original service ranking. The rank reversal problem or also known as the raking abnormality occurs in widely known MCDM approaches like Analytic Hierarchy Process (AHP), Technique for Order of Preference by Similarity to Ideal Solution (TOPSIS), Preference Ranking Organization METHod for Enrichment Evaluation (PROMETHEE), Performance Selection Index (PSI), ELimination and Choice Expressing REality (ELECTRE), etc., due to the relativity measurement [9]. The relativity measurement produce dependency that leads to rank reversal problem. As an initiative to address the rank reversal problem [10], this paper presents Grey CRiteria Importance through Intercriteria Correlation Parallel Analysis (GCRITICPA), a novel trust based service ranking approach which employs CRITICA to determine criteria weights and Grey Relational Analysis (GRA) to rank the trustworthy cloud service providers. A case study using Cloud Armor, a real-world trust feedback dataset demonstrates the flexibility and efficiency ranking of GCRITICPA over the existing MCDM methods like improved TOPSIS and Performance Selection Index (PSI) in terms of trustworthiness and consistent ranking. To the best of our knowledge, this is the first work to study and address the rank reversal phenomenon for cloud service selection with an illustrative example.

The rest of the paper is organized as follows: Sect. 2 describes the working of GCRITICPA. Section 3 presents a case study to analyze the performance of the proposed ranking approach. Section 4 concludes the paper.

Table 1. Related works

Authors	Technique Proposed	Dataset	Validation
Somu et al. [11]	Hypergraph – Binary Fruit Fly Optimization based service ranking Algorithm (HBFFOA) – identify suitable and trustworthy CSPs ✓ Hypergraph partitioning-identification of similar service providers ✓ Time-varying mapping function-credibility based trust assessment ✓ Helly property-select trustworthy service providers ✓ Binary fruit fly optimization-optimal service ranking	WSDream#2	Precision, stability, statistical test, and time complexity analysis
Nawaz et al. [12]	Cloud broker architecture – cloud service selection with respect to the change in User Preferences (Ups) over time ✓ Markov chain - generates a pattern based on change in priorities of user preferences ✓ Best Worst Method (BMW) – rank the cloud services	Case study-Amazon Elastic Compute (Amazon EC2) IaaS services with 4 services and 4 criteria	Pairwise comparison and convergence perspective
Araujo [13]	Novel decision-making with stochastic models considering availability, capacity-oriented availability (COA), reliability and cost requirements for cloud infrastructures ✓ Multi-criteria tool for Planning and Analysis of Cloud Environments (MiPACE) – consider customer service constraints ✓ TOPSIS – ranking cloud infrastructures	Case study	Mean Time To Failure (MTTF) & Mean Time To Repair (MTTR)
Soltani et al. [14]	Automatic cloud service selection system + TOPSIS – search based service selection with respect to the case-based reasoning	Case study – 7 cloud services and 5 attributes	Service recommendation prototype utilizing case-based recommendation approach
Yang et al. [15]	DEcision MAking Trial and Evaluation Laboratory (DEMATEl)-based Analytic Network Process (ANP) (DANP) – organize the weights of attributes with respect to their level of influence ✓ DEMATEL-construct an Influential Network Relationship Map (INRM) to correlate among influential networks ✓ ANP-determine attributes weights ✓ VIKOR (VIseKriterijumska Optimizacija I Kompromisno Resenje, it means Multicriteria Optimization and Compromise Solution) - estimates the sizes of the divergence among actual and desired service delivery performance	Case study-Taiwan's cloud application services (3 dimensions and 12 key factors)	Local weight and global weight based on DANP

(continued)

Table 1. (*continued*)

Authors	Technique Proposed	Dataset	Validation
Yadav and Goraya [16]	Two-way Ranking based Cloud Service Mapping – maps the Cloud Service Providers (CSPs) and Service Requesting Customers (SRCs) according to their QoS parameters ✓ Analytic Hierarchy Process (AHP) – ranking CSPs and Service Requesting Customers (SRCs)	Case Study- 3 service providers and 6 criteria	Sensitivity analysis
Ranjan Kumar et al. [17]	✓ AHP – significance (weights) of criteria ✓ TOPSIS – Ranking CSPs	Cloud Harmony	
Jatoth et al. [18, 19]	✓ Super-efficiency Data Envelopment Analysis (SDEA) – rank the Decision Making Units (DMUs) based on their performance ✓ Modified Data Envelopment Analysis (MDEA) – identify the preferred efficiency of cloud services using AHP	Case Study – 11 CSPs and 7 criteria	✓ Sensitivity analysis ✓ Adequacy change in alternatives (cloud services) ✓ Adequacy to support group decision making ✓ Uncertainty
	SELCLOUD – hybrid MCDM approach for cloud service selection ✓ AHP – determine criteria weights ✓ Extended Grey TOPSIS (EGTOPSIS) – rank the CSPs	Case study-19 cloud services and 5 QoS parameters	
Singh and Sidhu [20, 21]	✓ AHP – determine QoS parameters weights ✓ Improved TOPSIS – ranking CSPs ✓ PROMETHEE - Uncertainty	Cloud Armor & CloudHarmony	Trustworthiness, untrustworthiness, and uncertainty
Tripathi et al. [22]	Service Measurement Index (SMI) cloud framework - ANP for ranking cloud services	Case Study – 3 alternatives and 16 criteria	Sensitivity analysis
Radulescu and Radulescu [23]	Extended TOPSIS (E-TOPSIS)- ranking CSPs	Case study-10 CSPs and 3 criteria	TOPSIS
Liu et al. [24]	Multi Attribute Group Decision Making (MAGDM)-Cloud vendor selection ✓ Statistical Variance (SV) & Simple Additive Weighting (SAW) – compute attribute weights ✓ improved TOPSIS, and Delphi-AHP – determine the decision maker's weights ✓ Linguistic Weighted Arithmetic Averaging (LWAA) – rank the cloud vendors	Case study-4 alternatives, 3 attributes, and 4 decision makers	-
Shetty et al. [25]	REMBRANDT	Case study-4 cloud services and 6 QoS parameters	Rank reversal
Do and Kwang-Kyu [26]	Analytic Network Process – determine the weights, priorities of decision-making criteria, sub-criteria, and rank the cloud services	Case study – 6 cloud services and 3 attributes	-
Khan and Qamar [27]	Cloud Service Evaluation and Ranking Model (CSERM) (AHP and TOPSIS – ranking CSPs)	Case Study -10 alternatives and 7 criteria	-

(*continued*)

Table 1. (*continued*)

Authors	Technique Proposed	Dataset	Validation
Garg et al. [28]	Service Measurement Index (SMI) cloud framework -AHP for ranking cloud services	Case study - 3 CSPs & 6 attributes	Time complexity
Ghafori and Sarhadi [29]	ANP – criteria weights DEMATEL – rank the CSPs	Case Study-3 alternatives & 7 criteria	-
Petković et al. [30]	Performance Selection Index (PSI)	Case study 1-Machinability of materials Case study 2-selection of cutting fluids	Graph Theory and Matrix Approach (GTMA), TOPSIS, PROMETHEE

2 Proposed Methodology

Grey Relational Analysis (GRA) is popular over the state-of-the-art MCDM approaches since it has proven its efficiency in addressing the rank reversal problem [31]. Hence, this work presents GCRITICPA, a novel trust based service ranking approach for the identification trustworthy cloud service providers with consistent service ranking. The novelty of the proposed methodology (GCRITICPA) resides in hybridizing the CRiteria Importance Through Intercorrelation Criteria (CRITIC) to determine the objective weights of attributes based on initial evaluation matrix and Grey Relational Analysis for service ranking. None of the objective weight approaches are used to derive the weights from a given evaluation matrix i.e., before normalizing the evaluation matrix. The proposed framework comprises four phases namely, (i) Data representation phase – represents the set of alternatives (CSPs) and criteria (QoS parameters) in terms of preference matrix; (ii) Data pre-processing phase – eliminates the effect of various dimensions using target-based normalization technique; (iii) Weight computation phase – determine the importance of criteria through CRITIC method which utilizes the Pearson correlation method to find the correlation between the criteria and then compute the weights of each criteria; (iv) Ranking phase – The trustworthy cloud service providers are ranked based on the grey relational coefficient and grey relational degree (Fig. 1).

Step 1 – Data Representation: Determine the alternatives (CSPs) and criteria (QoS parameters). Construct a performance matrix (CSP_{m*n}) of order $m \, X \, n$, where m represents the CSPs (alternatives) and n represents the criteria (QoS parameters). The performance matrix (CSP_{m*n}) can be represented as follows (Table 2),

The Table 2 of performance matrix $(\mathfrak{p}_{ij})$, $P_{11}, P_{21}, \cdots, P_{mn}$ are the compliance values of n QoS parameters.

Step 2 – Data Normalization: Target-based normalization method is used to normalize the performance matrix for data pre-processing. In general, normalization eliminates the effect of different magnitudes and improves the precision of the proposed method. The resultant target-based normalized performance matrix $\mathfrak{U} = (\mathfrak{U}_{ij})_{m*n}$ can be obtained using Eq. (1).

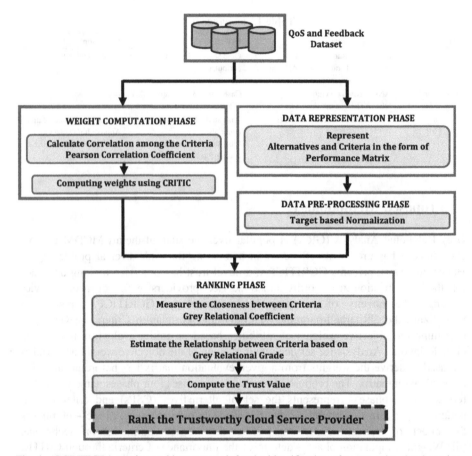

Fig. 1. GCRITICPA service ranking approach for the identification of trustworthy cloud service providers

Table 2. Performance matrix $\left(P_{ij} \right)$ of $CSP_{15 \times 9}$

CSPs	Criteria					
	A_1	A_2	$\cdots$	A_j	$\cdots$	A_n
CSP_1	P_{11}	P_{12}	$\cdots$	P_{12}	$\cdots$	P_{1n}
CSP_2	P_{21}	P_{22}	$\cdots$	P_{2j}	$\cdots$	P_{2n}
$\vdots$	$\vdots$	$\vdots$	$\ddots$	$\vdots$	$\ddots$	$\vdots$
CSP_i	P_{i1}	P_{i2}	$\cdots$	P_{ij}	$\cdots$	P_{in}
$\vdots$	$\vdots$	$\vdots$	$\ddots$	$\vdots$	$\ddots$	$\vdots$
CSP_m	P_{m1}	P_{m2}	$\cdots$	P_{mj}	$\cdots$	P_{mn}

$$\mathfrak{U}_{ij} = 1 - \frac{|T_j - P_{ij}|}{Max\left\{P_{ij}^{max}, T_j\right\} - Min\left\{P_{ij}^{min}, T_j\right\}} \tag{1}$$

Where, T_j represents the target value of the attributes. The target values of the attributes are calculated using Eq. (2)

$$T_j = \begin{cases} max_i \ x_{ij}, if \ j \in B, \\ min_i \ x_{ij}, \ if \ j \in C, \\ g_j, if \ j \in T \end{cases} \tag{2}$$

Where, B, C, and T represents the benefit, cost, and target-based attributes. The term g_j represents the goal value of target-based attribute.

Step 3 – Weight Computation: CRiteria Importance through Intercriteria Correlation (CRITIC) is used to compute the unbiased (statistical) weight of each criteria based on its correlation to all other criterion of a given performance matrix using Eqs. (3) and (4).

$$R_{jk} = \frac{\sum_{i=1}^{m}\left(P_{ij} - \overline{P_j}\right)\left(P_{ik} - \overline{P_k}\right)}{\sqrt{\sum_{i=1}^{m}\left(P_{ij} - \overline{P_j}\right)^2 \sum_{i=1}^{m}\left(P_{ik} - \overline{P_k}\right)^2}}; \forall j = 1, 2, \cdots, n; k = 1, 2, \cdots, n \tag{3}$$

Where, $m, \overline{P_j}, \overline{P_k}$ are the number of cloud service providers (alternatives) and the average values of criterion j and k respectively. The value of R_{jk} lies in the interval $[-1, +1]$ which indicates highly correlated criteria and the R_{jk} value nearest to zero represents no correlation.

$$\mathbb{W}_j = \frac{\Sigma_{k=1}^{n}(1-|R_{jk}|)}{\Sigma_{j=1}^{n}(\Sigma_{k=1}^{n}(1-|R_{jk}|))}; \forall \ j = 1, 2, \cdots, n; \ k = 1, 2, \cdots, m; \ k \neq j \tag{4}$$

Step 4 – Determine Reliable Ideal Solution (RIS) and Non-reliable Ideal Solution (NIS): The deviation sequence of alternatives are expressed by the computation of reliable ideal alternative (A_R) and non-reliable ideal alternative (A_N) for every criteria using Eqs. (5) and (6)

$$A_R = A_j^+ = \begin{cases} max_{1 \leq i \leq m} \ \mathfrak{U}_{ij}|j \in J^+ \ (\text{Benefit criteria}), \forall j = 1, 2, \cdots, n \\ min_{1 \leq i \leq m} \ \mathfrak{U}_{ij}|j \in J^- (\text{Cost criteria}), \forall j = 1, 2, \cdots, n \end{cases} \tag{5}$$

$$A_N = A_j^- = \begin{cases} min_{1 \leq i \leq m} \ \mathfrak{U}_{ij}|j \in J^+ \ (\text{Benefit criteria}), \forall j = 1, 2, \cdots, n \\ max_{1 \leq i \leq m} \ \mathfrak{U}_{ij}|j \in J^- (\text{Cost criteria}), \forall j = 1, 2, \cdots, n \end{cases} \tag{6}$$

Step 5– Grey Relational Coefficient (GRC) Computation: GRC express the relationship between the desired sequence (maximum compliance value of each criteria in $\mathfrak{U}_{ij}$) and comparable sequences (compliance values in $\mathfrak{U}_{ij}$). Thereby, GRC between the

i^{th} alternative and reliable ideal alternative and non-reliable ideal alternative about j^{th} QoS parameter is calculated according to Eq. (7) ("*" indicates "+" or "−")

$$\mathfrak{D}_{ij}^* = \frac{\min_i \min_j \left|A_j^* - \mathfrak{U}_{ij}\right| + \delta\left(\max_i \max_j \left|A_j^* - \mathfrak{U}_{ij}\right|\right)}{\left|A_j^* - \mathfrak{U}_{ij}\right| + \delta\left(\max_i \max_j \left|A_j^* - \mathfrak{U}_{ij}\right|\right)} \tag{7}$$

Where, δ is the distinguishing coefficient, $\delta \in [0, 1]$. The purpose of the distinguishing coefficient (δ) is to compress or expand the range of GRC. The default value of δ is 0.5 and is the widely accepted value [32]. Yiyo et al. [33] stated "after grey relational generating, $\max_i \max_j \left|A_j^* - \mathfrak{U}_{ij}\right| = 1$, and $\min_i \min_j \left|A_j^* - \mathfrak{U}_{ij}\right| = 0$".

Step 6 – Grey CRITIC Parallel Relational Grade (GCPRG) Computation: The CRITIC weightage value is assigned to the grey relational coefficient to compute GCPRG as defined in Eq. (8)

$$GCPRG_i^* = \mathfrak{N}_i^* = \sum_{j=1}^n (W_j * \mathfrak{D}_{ij}^*), \forall i = 1, 2, \cdots, m \tag{8}$$

where $GCPRG_i^*$ represents the $\mathfrak{N}_i^+$ and $\mathfrak{N}_i^-$ respectively

Step 7 – Cloud Service Ranking: Rank the cloud service providers with respect to the trustworthiness (T_i) as defined in Eq. (9).

$$T_i = \frac{\mathfrak{N}_i^*}{(\mathfrak{N}_i^* + (1 - \mathfrak{N}_i^*) + Var_i)} \tag{9}$$

Where, var_i represents the variance of the target-based normalized performance matrix $\left(\mathfrak{U}_{ij}\right)$.

3   Experimental Results – Case Study

This section presents a case study to validate the proposed approach (GCRITICPA) using a sample dataset (15 CSPs and 9 QoS attributes) extracted from Cloud Armor, a research project at the University of Adelaide [34]. It contains 10,080 QoS feedbacks ranging from 1 (insignificant feedback score) and 5 (significant feedback score) provided by approximately 7000 consumers for 114 real world cloud services during the time period from July 2003 to June 2012. This case study includes trust feedback values given for fifteen CSPs namely, Backupgenie (*CSP1*), Bluehost (*CSP2*), Carbonite (*CSP3*), Elephantdrive (*CSP4*), Go Daddy (*CSP5*), ibackup (*CSP6*), idrive (*CSP7*), Justcloud (*CSP8*), Keepit (*CSP9*), Livedrive (*CSP10*), Mozy (*CSP11*), MyPCBackup (*CSP12*), sos-online-backup (*CSP13*), SugarSync (*CSP14*), and yousendit-online-backup (*CSP15*) over nine QoS attributes, namely availability (*Av*), response time (*Rt*), price (*Pr*), speed (*Sp*), storage space (*Ss*), features (*fe*), ease of use (*Eu*), technical support (*Ts*), customer service (*Cs*).

The performance matrix $(p_{ij})_{15\times9}$ of fifteen CSPs and nine QoS attributes are presented in Table 3. Further, the quantitative values of the performance matrix (for CSPs) was normalized using target-based normalization technique for the construction of the normalized performance matrix (Eq. (1)).

Table 3. Performance matrix $CSP_{15\times9}$

CSPs	Criteria								
	Av	Rt	Pr	Sp	Ss	Fe	Eu	Ts	Cs
CSP1	5	5	5	3	5	5	5	5	5
CSP2	5	5	5	4	4	5	5	5	5
CSP3	3	3	3	4	5	2	2	2	2
CSP4	5	4	4	4	3	4	5	5	5
CSP5	5	5	5	5	4	5	5	5	5
CSP6	5	5	5	5	5	5	5	5	5
CSP7	4	4	4	4	4	5	5	4	4
CSP8	5	5	5	5	5	5	5	4	4
CSP9	3	4	4	3	4	3	5	4	4
CSP10	5	4	4	1	4	4	1	1	1
CSP11	2	3	2	3	3	3	3	2	3
CSP12	5	4	4	5	5	5	5	4	4
CSP13	3	3	3	3	2	3	3	4	3
CSP14	5	5	4	5	5	5	5	5	5
CSP15	5	5	4	5	4	4	5	5	5

The preference matrix was normalized using target-based normalization technique for the construction of the normalized performance matrix (Eq. (1)). The significance of each attribute (weights) was determined using CRITIC, an objective weight method (using Eqs. (3) and (4)) from the performance matrix. Table 4 presents the normalized performance matrix and the weights of attributes.

The grey relational coefficient was determined by comparing the reference sequence with the comparable sequence to represent the closeness among them (Eq. (7)). Table 5 presents the values of the GRC, and GCPRG (Eqs. (7) and (8)).

Table 6 presents the trustworthiness and ranking of cloud services (Eq. (8))

The service ranking of the 15 CSPS based on their trustworthiness is as follows:

ibackup > *SugarSync* > *justcloud* > *GoDaddy* > *Backupgenie*
> *Bluehost* > *yousendit* − *online* − *backup* > *MyPCBackup* > *Elephantdrive*
> *idrive* > *keepit* > *Carbonite* > *livedrive* > *sos* − *online* − *backup* > *Mozy*

The trustworthy cloud service provider is "ibackup (CSP6)". The CSP5 (GoDaddy) drop to the fourth place due to the summation of three values (GCPRG, variance and the value of 1-GCPRG), which leads to the consistent service ranking. The accuracy

Table 4. Normalized performance matrix $NCSP_{15 \times 9}$

CSPs	Av	Rt	Pr	Sp	Ss	Fe	Eu	Ts	Cs
CSP1	1.000	1.000	1.000	0.500	1.000	1.000	1.000	1.000	1.000
CSP2	1.000	1.000	1.000	0.750	0.667	1.000	1.000	1.000	1.000
CSP3	0.333	0.000	0.333	0.750	1.000	0.000	0.250	0.250	0.250
CSP4	1.000	0.500	0.667	0.750	0.333	0.667	1.000	1.000	1.000
CSP5	1.000	1.000	1.000	1.000	0.667	1.000	1.000	1.000	1.000
CSP6	1.000	1.000	1.000	1.000	1.000	1.000	1.000	1.000	1.000
CSP7	0.667	0.500	0.667	0.750	0.667	1.000	1.000	0.750	0.750
CSP8	1.000	1.000	1.000	1.000	1.000	1.000	1.000	0.750	0.750
CSP9	0.333	0.500	0.667	0.500	0.667	0.333	1.000	0.750	0.750
CSP10	1.000	0.500	0.667	0.000	0.667	0.667	0.000	0.000	0.000
CSP11	0.000	0.000	0.000	0.500	0.333	0.333	0.500	0.250	0.500
CSP12	1.000	0.500	0.667	1.000	1.000	1.000	1.000	0.750	0.750
CSP13	0.333	0.000	0.333	0.500	0.000	0.333	0.500	0.750	0.500
CSP14	1.000	1.000	0.667	1.000	1.000	1.000	1.000	1.000	1.000
CSP15	1.000	1.000	0.667	1.000	0.667	0.667	1.000	1.000	1.000
W_j	**0.1119**	**0.0829**	**0.0985**	**0.1346**	**0.1748**	**0.0994**	**0.0985**	**0.1019**	**0.0976**

Table 5. Grey relational coefficient and grey relational grade values

CSPs	Av	Rt	Pr	Sp	Ss	Fe	Eu	Ts	Cs	GCPRG
CSP1	0.1119	0.0829	0.0985	0.0673	0.1748	0.0994	0.0985	0.1019	0.0976	0.9328
CSP2	0.1119	0.0829	0.0985	0.0897	0.1049	0.0994	0.0985	0.1019	0.0976	0.8853
CSP3	0.0480	0.0276	0.0422	0.0897	0.1748	0.0331	0.0394	0.0408	0.0390	0.5347
CSP4	0.1119	0.0415	0.0591	0.0897	0.0749	0.0596	0.0985	0.1019	0.0976	0.7347
CSP5	0.1119	0.0829	0.0985	0.1346	0.1049	0.0994	0.0985	0.1019	0.0976	0.9302
CSP6	0.1119	0.0829	0.0985	0.1346	0.1748	0.0994	0.0985	0.1019	0.0976	1.0000
CSP7	0.0671	0.0415	0.0591	0.0897	0.1049	0.0994	0.0985	0.0679	0.0651	0.6932
CSP8	0.1119	0.0829	0.0985	0.1346	0.1748	0.0994	0.0985	0.0679	0.0651	0.9336
CSP9	0.0480	0.0415	0.0591	0.0673	0.1049	0.0426	0.0985	0.0679	0.0651	0.5948
CSP10	0.1119	0.0415	0.0591	0.0449	0.1049	0.0596	0.0328	0.0340	0.0325	0.5212
CSP11	0.0373	0.0276	0.0328	0.0673	0.0749	0.0426	0.0493	0.0408	0.0488	0.4214
CSP12	0.1119	0.0415	0.0591	0.1346	0.1748	0.0994	0.0985	0.0679	0.0651	0.8528
CSP13	0.0480	0.0276	0.0422	0.0673	0.0583	0.0426	0.0493	0.0679	0.0488	0.4520
CSP14	0.1119	0.0829	0.0591	0.1346	0.1748	0.0994	0.0985	0.1019	0.0976	0.9607
CSP15	0.1119	0.0829	0.0591	0.1346	0.1049	0.0596	0.0985	0.1019	0.0976	0.8510

and flexibility of the proposed trust based service ranking approach (GCRITICPA) was compared with the state-of-the-art MCDM approaches like Improved TOPSIS [19] and PSI. From Table 7, it is clear that trust based service ranking of GCRITICPA has high degree of similarity with the Spearman rank correlation coefficient and the ranking provided by improved TOPSIS and PSI.

Table 6. Trustworthiness and ranking of cloud service

CSPs	T_i	Rank
CSP1	0.9076	5
CSP2	0.8705	6
CSP3	0.4829	12
CSP4	0.6918	9
CSP5	0.9188	4
CSP6	1.0001	1
CSP7	0.6756	10
CSP8	0.9224	3
CSP9	0.5682	11
CSP10	0.4521	13
CSP11	0.4020	15
CSP12	0.8231	8
CSP13	0.4268	14
CSP14	0.9490	2
CSP15	0.8280	7

Table 7. Trust based service ranking – Improved TOPSIS, PSI and GCRITICPA

MCDM approach	Ranking
Improved TOPSIS	$CSP6 > CSP14 > CSP8 > CSP5 > CSP12 > CSP1 > CSP2$ $> CSP15 > CSP7 > CSP4 > CSP9 > CSP3 > CSP10 > CSP13 > CSP11$
PSI	$CSP6 > CSP14 > CSP8 > CSP5 > CSP1 > CSP2 > CSP15 > CSP12$ $> CSP7 > CSP4 > CSP9 > CSP3 > CSP10 > CSP13 > CSP11$
GCRITICPA	$CSP6 > CSP14 > CSP8 > CSP5 > CSP1 > CSP2 > CSP15 > CSP12$ $> CSP4 > CSP7 > CSP9 > CSP3 > CSP10 > CSP13 > CSP11$

From Fig. 2 we observe that, the GCRITICPA method is moderately correlated with PSI (96%) and weakly correlated with Improved TOPSIS method (75%).

Further, the proposed service ranking approach addresses the rank reversal problem in cloud service selection. For instance, if the worst alternative (CSP) Mozy (CSP11) is removed from the considered set of alternatives (CSPs) (Table 6), the ranking provided by the Improved TOPSIS was found to be inconsistent since the ranking order is altered. For the same, the ranking provided by GCRITICPA reflects that the proposed service ranking approach addresses the rank reversal problem. For instance, the worst CSP i.e., CSP11 (Mozy) was removed from the list of CSPs. Then the service ranking of 14 CSPs is $\mathbf{CSP6} > CSP14 > CSP8 > CSP5 > CSP1 > CSP2 > CSP15 > CSP12 > CSP4 > CSP7 > CSP9 > CSP3 > CSP10 > CSP13$. Similarly, if the

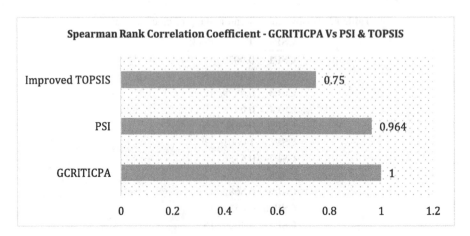

Fig. 2. Spearman's Rank Correlation Coefficient

existing cloud service provider, say, CSP2 was added to the list of CSPs then the ranking of 16 CSPs is **CSP6** $> CSP14 > CSP8 > CSP5 > CSP1 > CSP2 \cong CSP2 > CSP15 > CSP12 > CSP4 > CSP7 > CSP9 > CSP3 > CSP10 > CSP13 > CSP11$.

4   Conclusions

Service selection, a significant research topic for the various researchers and industry professionals in service oriented environments. Specifically, service selection has gained its attention in the cloud marketplace due to intrinsic nature of cloud computing. Hence, this work presents GCRITICPA, a novel trust based service ranking approach that employs CRITIC (importance of weights) and Grey Relational Analysis (ranking) for the identification of the trustworthy CSPs. The efficiency of the proposed service ranking approach was demonstrated using a case study with 15 CSPs and 9 criteria obtained from CloudArmor, a real world trust feedback dataset in terms of trustworthiness and consistent ranking.

Acknowledgements. This work was supported by The Council for Scientific and Industrial Research and The Department of Science and Technology, India (Grant No: CSIR - SRF Fellowship/143345/2K17/1, DST/INSPIRE Fellowship/2013/963, CSIR - SRF Fellowship/ 143404/2K15/1, and SR/FST/ETI- 349/2013).

References

1. Armbrust, M.Z.M., et al.: A view of cloud computing. Commun. ACM **53**(4), 50–58 (2010)
2. Liang, H., Du, Y.: Dynamic service selection with QoS constraints and inter-service correlations using cooperative coevolution. Futur. Gener. Comput. Syst. **76**, 119–135 (2017)

3. Somu, N., Kirthivasan, K., Shankar Sriram, V.S.: A rough set-based hypergraph trust measure parameter selection technique for cloud service selection. J. Supercomput. **73**(10), 4535–4559 (2017)
4. Thampi, S.M., Atrey, P.K., Bhargava, B.: Managing Trust in Cyberspace. CRC Press, Boca Raton (2013)
5. Qu, L.: Credible Service Selection in Cloud Environments, Macquarie University (2016)
6. Tang, M., Dai, X., Liu, J., Chen, J.: Towards a trust evaluation middleware for cloud service selection. Futur. Gener. Comput. Syst. **74**, 302–312 (2016)
7. Alabool, H., Kamil, A., Arshad, N., Alarabiat, D.: Cloud service evaluation method-based multi-criteria decision-making: a systematic literature review. J. Syst. Softw. **139**, 161–188 (2018)
8. Belton, V., Gear, T.: On a short-coming of Saaty's method of analytic hierarchies. Omega **11** (3), 228–230 (1983)
9. Bendaoud, F., Didi, F., Abdennebi, M.: A modified-SAW for network selection in heterogeneous wireless networks. ECTI Trans. Electr. Eng. Electron. Commun. **15**(2), 8–17 (2017)
10. Akshya Kaveri, B., Gireesha, O., Somu, N., Gauthama Raman, M.R., Shankar Sriram, V.S.: E-FPROMETHEE: an entropy based fuzzy multi criteria decision making service ranking approach for cloud service selection. In: venkataramani, gp, Sankaranarayanan, K., Mukherjee, S., Arputharaj, K., Sankara Narayanan, S. (eds.) ICIIT 2017. CCIS, vol. 808, pp. 224–238. Springer, Singapore (2018). https://doi.org/10.1007/978-981-10-7635-0_17
11. Somu, N., Gauthama Raman, M.R., Kannan, K., Shankar Sriram, V.S.: A trust centric optimal service ranking approach for cloud service selection. Futur. Gener. Comput. Syst. **86**, 234–252 (2018)
12. Nawaz, F., Asadabadi, M.R., Janjua, N.K., Hussain, O.K., Chang, E., Saberi, M.: An MCDM method for cloud service selection using a Markov chain and the best-worst method. Knowl. Based Syst. **159**, 120–131 (2018)
13. Araujo, J., Maciel, P., Andrade, E., Callou, G., Alves, V., Cunha, P.: Decision making in cloud environments: an approach based on multiple-criteria decision analysis and stochastic models. J. Cloud Comput. **7**(1), 1–19 (2018)
14. Soltani, S., Martin, P., Elgazzar, K.: A hybrid approach to automatic IaaS service selection. J. Cloud Comput. **7**(12), 1–18 (2018)
15. Yang, M.H., Su, C.H., Wang, W.C.: Use of hybrid MCDM model in evaluation for cloud service application improvement. EURASIP J. Wirel. Commun. Netw. **98**, 1–8 (2018)
16. Yadav, N., Goraya, M.S., Goraya, M.S.: Two-way ranking based service mapping in cloud environment. Futur. Gener. Comput. Syst. **81**, 53–66 (2017)
17. Ranjan, R., Siba, K., Chiranjeev, M.: A novel framework for cloud service evaluation and selection using hybrid MCDM methods. Arab. J. Sci. Eng. **43**, 1–16 (2017)
18. Jatoth, C., Gangadharan, G.R., Fiore, U.: Evaluating the efficiency of cloud services using modified data envelopment analysis and modified super-efficiency data envelopment analysis. Soft. Comput. **21**(23), 7221–7234 (2017)
19. Jatoth, C., Gangadharan, G.R., Fiore, U., Buyya, R.: SELCLOUD: a hybrid multi-criteria decision-making model for selection of cloud services. Soft Comput., 1–15 (2018). https://doi.org/10.1007/s00500-018-3120-2
20. Singh, S., Sidhu, J.: Compliance-based multi-dimensional trust evaluation system for determining trustworthiness of cloud service Providers. Futur. Gener. Comput. Syst. **67**, 109–132 (2017)
21. Sidhu, J., Singh, S.: Design and comparative analysis of MCDM-based multi-dimensional trust evaluation schemes for determining trustworthiness of cloud service providers. J. Grid Comput. **15**, 197–218 (2017)

22. Tripathi, A.: Integration of analytic network process with service measurement index framework for cloud service provider selection, pp. 1–16 (2017)
23. Rădulescu, C.Z., Rădulescu, I.C.: An extended TOPSIS approach for ranking cloud service providers. Stud. Inform. Control **26**, 183–192 (2017)
24. Liu, S., Chan, F.T.S., Ran, W.: Decision making for the selection of cloud vendor: an improved approach under group decision-making with integrated weights and objective/subjective attributes. Expert Syst. Appl. **55**, 37–47 (2016)
25. Shetty, J., D'Mello, D.A.: Quality of service driven cloud service ranking and selection algorithm using REMBRANDT approach. In: 2015 International Conference on Smart Technologies and Management for Computing, Communication, Controls, Energy and Materials (ICSTM). IEEE (2015)
26. Do, C.B., Kwang-Kyu, S.: A cloud service selection model based on analytic network process. Indian J. Sci. Technol. **8**, 1–5 (2015). http://dx.doi.org/10.17485/ijst/2015/v8i18/77721
27. Khan, M.Z., Qamar, U.: Towards service evaluation and ranking model for cloud infrastructure selection. In: Ubiquitous Intelligence and Computing and 2015 IEEE 12th International Conference on Autonomic and Trusted Computing and 2015 IEEE 15th International Conference on Scalable Computing and Communications and Its Associated Workshops (UIC-ATC-ScalCom) (2015)
28. Garg, S.K., Versteeg, S., Buyya, R.: A framework for ranking of cloud computing services. Future Gener. Comput. Syst. **29**(4), 1012–1023 (2013)
29. Ghafori, V., Sarhadi, R.M.: Best cloud provider selection using integrated ANP-DEMATEL and prioritizing SMI attribute. Int. J. Comput. Appl. **71**(16), 18–25 (2013)
30. Petković, D., Madić, M., Radovanović, M., Gečevska, V.: Application of the performance selection index method for solving machining MCDM problems. Facta Universitatis Ser. Mech. Eng. **15**(1), 97–106 (2017)
31. Huszák, Á., Imre, S.: Eliminating rank reversal phenomenon in GRA – based network selection method. In: IEEE International Conference on Communications, pp. 1–6 (2010)
32. Chin, P.F.: Manufacturing process optimization for wear property of fiber reinforced polybutylene terephthalate composites with grey relational analysis. Wear **254**(3–4), 298–306 (2003)
33. Yiyo, K., Taho, Y., Guan, W.H.: The use of a grey-based Taguchi method for optimizing multi-response simulation problems. Eng. Optim. **40**(6), 517–528 (2008)
34. Cloud Armor Project. http://cs.adelaide.edu.au/~cloudarmor/. Available 31 August 2015. Accessed 22 Sept 2018 at 2:00 p.m

Cloud Enabled Intrusion Detector and Alerter Using Improved Deep Learning Technique

K. Kanagaraj[1(✉)], S. Swamynathan[2], and A. Karthikeyan[1]

[1] MEPCO Schlenk Engineering College, Sivakasi, Tamilnadu, India
kanagaraj@mepcoeng.ac.in, karthikeyana97@gmail.com
[2] College of Engineering, Guindy, Anna University, Chennai, Tamilnadu, India
swamyns@annauniv.edu

Abstract. Deep learning is a popular machine learning technique, used in variety of applications like autonomous vehicles, aerospace and medical research. Wireless Integrated Network Sensors (WINS) is an architecture that provide continuous monitoring and control of an environment with high accuracy and low power consumption. The amalgamation of deep learning and WINS can provide an effective system to monitor the remote environment. In this paper a system is designed using WINS and Improved Convolution Neural Network (ICNN). This can be used to create a virtual wall across the border of the county. This virtual wall is made up of sensors and cameras that are placed at regular intervals. The images of the suspected intruders are captured and are classified by ICNN. The processed image is stored in the fire base cloud which in turn will alert the country border security authorities by sending a message to them using MQTT protocol. This system is effective and can easily identify the intruders instantaneously. The experimental results shows that the improved convolution algorithm performs image classification better and faster than the traditional convolution algorithms.

Keywords: Border security · Cloud monitoring · Deep learning
Sensor network

1 Introduction

Wireless Integrated Network Sensors (WINS) is an outcome of the latest technological development in the form of Internet of Things. The WINS contains sensors, actuator, and processing nodes. It consists of a collection of sensors and devices called as nodes. All the nodes should be made in such a way that they are independent and able to communicate with other nodes seamlessly. These sensors should be placed at regular intervals and should be linked using wireless sensors. Care should be taken to enable that the nodes to work with less power. The WINS can be applied in areas like aerospace, environment monitoring and security. This paper deals with the application of WINS network in border security.

Deep Learning is an interesting area of research. It is a part of machine learning that provide artificial Intelligence to machines. Convolutional Neural Networks (CNN) is a deep learning technique that is widely used for image classification. The automatic

L. Akoglu et al. (Eds.): ICIIT 2018, CCIS 941, pp. 17–29, 2019.
https://doi.org/10.1007/978-981-13-3582-2_2

detection and classification of images without human supervision is the unique advantage of CNNs when compared to other competent methods. The CNN is self dependent and learns the unique features of different classes of image by itself using max pooling or average pooling. The option to adopt max pooing or average pooling is based on the application and the nature of the input images. CNN has been recognized as an effective approach for image recognition and classification problems.

Firebase Cloud Messaging (FCM) is a multiplatform messaging tool that is offered as free of cost. The FCM adopts the publish/subscribe model. Messages to the firebase cloud can be uploaded using any authorized application. All the clients subscribed to it will automatically receive message from the FCM when a new message is uploaded. The clients can be any mobile application or a web application. It is an instant messaging platform, by which we can deliver messages with a maximum size of 4 KB. In this paper FCM is used to store details about the intruders in the cloud.

Message Queue Telemetry Transport (MQTT) is a lightweight messaging protocol. It provides a simple and easy mechanism for the users to exchange message in a simple way. It follows the publish/subscribe model which enable the communication between machines and devices easier. The simplicity and power of MQTT is very well utilized by the IoT. If we push the output to a repository, all the clients who have subscribed to it can receive the message automatically. The MQTT protocol is a good choice for wireless networks having constraints with respect to network and power. In this paper MQTT is used to send message and alert the officers and security personnel, if an intruder is found.

2 Literature Review

The need for WINS is to develop a scalable, low cost, sensor network environment. In WINS the sensor data will be sent to the user at low bit rate having low power transceivers. Incessant sensor signal monitoring must be provided to secure an environment. Conventional methods for establishing a sensor network may require extensive cable installation and high network bandwidth. As WINS performs processing at source, the burden on communication system components, networks, and human resources are significantly reduced.

The use and applications of wireless sensors in a variety of applications was studied and presented by Pottie and Kaiser [1]. They have described the physical ideologies that lead to consideration of dense sensor networks and also outlined, how energy and bandwidth constraints demand a distributed and layered signal processing architecture. Further the need for self-organization and reconfiguration are discussed. Also they have discussed how WINS nodes can best be embedded in the Internet. They have also designed a prototype platform that enables several functions including remote Internet control, and analysis of sensor network operation.

Vardhan et al. [2] have explored enabling technology advances of WINS used in mission and flight system applications. A complete set of technologies, from new MEMS devices through information technology were also outlined. They also developed GlobalWINS new WINS generation technology. The use of WINS in border security was studied by Shire et al. [3]. They have discussed the importance of using

WINS in the border instead of deploying human warriors. They have used spectrum analysed for detecting intruders.

The use of convolution neural network in image classification was studied by Yim et al. [4]. They adopted a method that combines the features of several layers in the CNN model. Moreover, to extract the features from different layers they have used the previously trained images. A survey of the use of CNN in crowd analysis was done by Tripathi et al. [5]. They have done a survey of the popular software tools for CNN in the recent years. This survey presents in detail the attributes of CNN with special emphasis on optimization methods that have been utilized in CNN based methods. They have also presented a review of the fundamental and innovative methodologies, both conventional and latest methods of CNN, developed in the recent past.

Anusae and Vyas [6] have presented a training algorithm that reduce the complete retraining of any neural network architecture used for visual pattern recognition problems. They have also investigated the performance of convolutional neural network (CNN) architecture for a face recognition task under transfer learning. The algorithm may be used for enhancing the utility of machine learning software by providing researchers with an approach that can reduce the training time under transfer learning. The box fusion method proposed by Cao et al. [7] is a new approach to detect objects like aircrafts in natural image. The approach also works well in complex sensing areas. Ding et al. [8] have proposed a multi-scale representation and various combinations of improved schemes to enhance the structure of the base VGG16-Net for improving the precision to reduce the test-time and memory requirements. The results shows that the improved network structure can detect objects in satellite optical remote sensing images more accurately and efficiently.

The features of Firebase cloud and the phases to integrate the FCM with our application was discussed in detail by Srivastava et al. [9]. They have also outlined the procedure to send and receive messages using FCM and Google Cloud Messaging (GCM). A research paper published by Soni et al. [10] describes the importance of MQTT in IoT. They have defined the structure of MQTT messaging and the various domain where it can be used. The different types of brokers used in MQTT and the current issues and future trends were also discussion in detail. del Rio et al. [14] have proposed an algorithm to secure the cross border traffic using a WINS network. Nunamaker Jr. et al. [15] have generated a prototype system capable of augmenting human screeners in detecting deception and assessing intent at border entry points.

Kendall [13] have proposed a technique to control the cross border traffic without compromising the security.

Alom et al. [16] have proposed a new DCNN model called the Inception Recurrent Residual Convolutional Neural Network (IRRCNN), that utilizes the power of the Recurrent Convolutional Neural Network (RCNN), the Inception network, and the Residual network. This approach improves the recognition accuracy of the Inception-residual network with same number of network parameter.

3   Intrusion Detection Systems Using ICNN

The intrusion detection system using the improved convolution neural network (ICNN) proposed in this paper consists of a network of connected devices and sensors, that act as a wall to protect the borders of the country. The architecture is shown in Fig. 1.

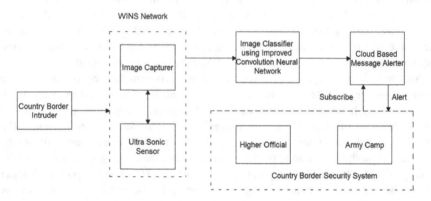

Fig. 1. Cloud Enabled Country Border Intruder Detection System using ICNN

All the sensors are connected using WINS and the captured data is processed using ICNN which is a powerful processing algorithm that can process and classify the image with 25% of speed than the traditional algorithms. This will help the security personnel to act immediately.

3.1   Wireless Integrated Network Sensors (WINS)

WINS was realized in the year 1993 under Defence advance research project agency (DARPA) in USA. In 1998, WINS NG was introduced for wide range of applications. WINS architecture includes sensor, data converter, signal processing, and control functions. The micro power components operate continuously for event recognition, while the network interface operates at low duty cycle. WINS performs contiguous sensing, signal processing for event detection, local control of actuators, event classification, and communication at low power. WINS nodes are distributed at high density in an environment to be monitored. WINS node data is transferred over the asymmetric wireless link to an end user.

If a stranger enters the border, his foot-steps will generate acoustic signals. It can be detected as a characteristic feature in the sensing node. The WINS architecture used in this paper is made up of sensors, microcontrollers, alarms and cameras. They are classified into three types of nodes. Sensing nodes, alarming node and capturing node. The sensed data is processed by the ICN algorithm and if an intruder is found, the data is sent to the fire base cloud. From the fire base cloud [11], the message is forwarded to those who are subscribed to it. From the cloud, the information about the intruder is passed on to the authorities using MQTT [12] protocol for appropriate action.

3.1.1 Sensing Node

The sensing node uses the ultrasonic sensor to detect the intrusion of an object in the border. The node is made up of ultrasonic sensor with Raspberry pi 3. Both of them are connected using wireless interface. A model of the sensing node is shown in Fig. 2.

Fig. 2. Components of sensing node

Fig. 3. Components of a capturing node

3.1.2 Capturing Node

The capturing node captures the image of the intruder. The capturing node will remain idle until it gets instruction from the sensing node. The sensing node consists of a camera and a Raspberry pi card, as shown in Fig. 3. Upon activation it captures the image. The captured image is processed by the improved convolution algorithm. Tensor flow 1.0 (Library for ML-python 3.0) is used for image classification. After processing, if the image is identified as a human then it stores the image to the fire base cloud using node.js push notification feature.

3.1.3 Alarm Node

The alarm node sounds the buzzer if an intruder is found. It is activated by the camera node. The buzzer sound will alert the security personnel near to the location to take the required action. The alarm node consists of a buzzer and a Raspberry pi card.

3.1.4 Packing of Sensor Nodes

The sensor node must be packed with acrylic material to save the node from extreme heat, light, snow or wind. While packing the nodes care should be given in such a way that the quality of signal is not affected.

3.2 Convolution Neural Network

CNNs are biologically inspired models introduced by research by D. H. Hubel and T. N. Wiesel. They proposed a clarification for the way in which mammals visually recognize the world around them using a layered architecture of neurons in the brain, and this in turn inspired engineers to develop similar pattern recognition mechanisms in computer vision. Convolution neural network consists of a multilayer perceptron that can process images with two and three dimensions. The CNN is made up of an input layer, convolution layer, sample layer and output layer. However, in a deep network architecture, the convolution layer and sample layer can have multiple levels. Every layer of the CNN is designed to process the image in its adjacent layer. Each layer can also perform the convolution and deconvolution if required. The convolution layer is composed of a trainable filter fx and a bias bx. In sampling, n pixels of each neighborhood values are pooled to generate a pixel, then by scalar weighting $Wx + 1$

weighted, add a bias bx + 1, and then using an activation function, generate a narrow n times feature map Sx + 1.

The key technology of CNN is the local receptive field, sharing of weights, sub sampling by time or space, that are useful to extract various feature and reduce the size of the training parameters. The neural network is designed to learn in parallel to reduce the complexities. Adopting sub sampling structure by time or space, can achieve some degree of robustness, scale and deformation. Input information and network topology can be a very good match. It has unique advantages in speech recognition and image processing.

3.2.1 Existing Convolution Neural Network (CNN)

CNN algorithms require experience in architecture design, and need to debug unceasingly in the practical application. This section explains how a grey image with an initial size of 96 × 96 is converted into an image of size 32 × 32 pixels. Figure 4 shows a CNN with seven layers depth.

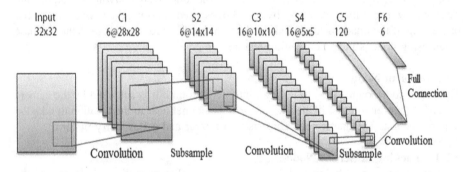

Fig. 4. Architecture of Convolution Neural Network

The C1 layer for convolution has 6 convolution kernels, and the size of each convolution kernels is 5 × 5, that can generate six feature map, that contains 784 neurons. In this layer 6 × (5 × 5 + 1) = 156 factors are trained. S1 layer for sub sampling, extracts six feature map. Each feature map requires 14 * 14 = 196 neurons. The sub sampling window is 2 × 2 matrix, sub sampling step size is 1, so the S1 layer contains 6 × 196 × (2 × 2 + 1) = 5880 connections. Every feature map in the S1 layer contains a weight and bias, so a total of 12 parameters can be trained in S1 layer. C2 is convolution layer, containing 16 feature graph, each feature graph contains (14 − 5 + 1) (14 − 5 + 1) = 100 neurons and adopts full connection. This process is repeated until all the features are extracted and the image is flattened.

3.2.2 Residual Convolution Neural Network (RCNN)

Deep convolutional neural networks have contributed for tremendous developments in image classification. By nature the deep networks integrate several features and classifiers in an end-to-end multilayer fashion. The levels of features can be enriched by the depth of the network. It is understood that the network depth is essential for image recognition what may go up to 20 layers. The deeper network has higher training error, and thus test error. Similar phenomena on ImageNet is greatly benefited from very deep models. Does the significance of depth contributes for learning is a big question? The use of intermediate layers enable networks with tens of layers to start converging for stochastic gradient descent (SGD) with backpropagation [17]. When deeper networks are able to start converging, a degradation problem has been exposed. When the network depth increases the accuracy gets saturated and then degrades rapidly. Unexpectedly, such degradation is not caused by overfitting, and adding more layers to a suitably deep model leads to higher training error. The degradation of training accuracy indicates that not all systems are similarly easy to optimize.

3.2.3 Improved Convolution Neural Network (ICNN)

The improved convolution algorithm is able to classify the images with great speed and accuracy than the traditional convolution algorithms. The primary purpose of convolution in case of a CNNs is to extract features from the input image. Convolution preserves the spatial relationship between pixels by learning image features using small squares of input data. Every image can be considered as a matrix of pixel values. However, it remains to be seen if these computational mechanisms of convolutional neural networks are similar to the computation mechanisms occurring in the primate visual system, convolution operation, and shared weights and pooling/subsampling. Every color image is composed of three primary colors red, green and blue (RGB). All these colors need to be processed separately. Figure 5 shows the feature extraction from an input image having Red, Green and Blue channels.

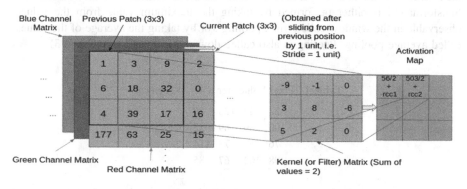

Fig. 5. Channel Matrix and Filtering process (Color figure online)

A convolution is an orderly procedure where two sources of information are intertwined. A kernel (also called a filter) is a smaller-sized matrix in comparison to the input dimensions of the image that consists of real valued entries. Kernels are then convolved with the input volume to obtain so-called activation maps Activation maps indicate activated regions. The activated regions features specific to the kernel have been detected in the input. The real values of the kernel matrix change with each learning iteration over the training set, indicating that the network is learning to identify which regions are of significance for extracting features from the data.

We compute the dot product between the kernel and the input matrix. The convolved value obtained by summing the resultant terms from the dot product forms a single entry in the activation matrix. The patch selection is then slided (towards the right, or downwards when the boundary of the matrix is reached) by a certain amount called the 'stride' value, and the process is repeated till the entire input image has been processed. The process is carried out for all colour channels.

Instead of connecting each neuron to all possible pixels, we specify a 2 dimensional region called the 'receptive field'. For a size of 5×5 units extending to the entire depth of the input ($5 \times 5 \times 3$ for a 3 colour channel input), within which the encompassed pixels are fully connected to the neural network's input layer. It's over these small regions that the network layer cross-sections, each consisting of several neurons called 'depth columns' that operate and produce the activation map.

When convolution is applied to images, two dimensions are required. The two dimension are the width and the height of the image. When these buckets of information are mixed together, the first bucket is the input image, which has a total of three matrices of pixels. One matrix each for the red, blue and green color channels. A pixel consists of an integer value between 0 and 255 in each color channel. The second bucket is the convolution kernel, a single matrix of floating point numbers where the pattern and the size of the numbers can be thought of as a recipe for how to intertwine the input image with the kernel in the convolution operation. The output of the kernel is the altered image which is often called a feature map in deep learning. There will be one feature map for every color channel.

Pooling reduces the spatial dimensions (Width $\times$ Height) of the input volume for the next convolutional Layer. It does not affect the depth dimension of the volume. The transformation is either performed by taking the maximum value from the values observable in the window called 'max pooling', or by taking the average of the values called average pooling. Pooling is also called down sampling (Tables 2 and 3).

Table 1. Pixel Values for a 4×4 segment

23	44	123	56
100	89	69	32
76	56	78	114
98	102	67	85

Table 2. Max Pooling for values in Table 1.

100	123
102	113

Table 3. Average Pooling for values in Table 1.

67	70
83	86

In the proposed algorithm, both max pooling and average pooling are used for down sampling the feature map. Instead of taking max pooling in all layers, we take max pool for first few layers and then average pool for rest of the layers. In first few layers we extract shapes and edges in the image, every weights in the feature map not much important. So, we apply max pool for the feature map. In any picture representation of feature map extraction, final layer captures every features of the people face. If we take max pool, some of the weights (features) will be neglected. So we have to take average pool in that layer. The pictorial representation of the ICNN algorithm is shown in Fig. 6.

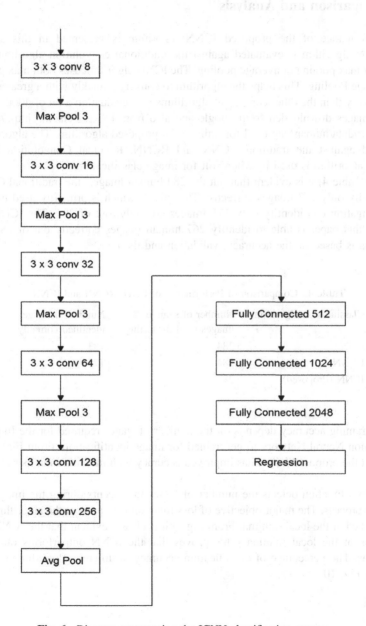

Fig. 6. Diagram representing the ICNN classification process

The general CNN uses Max pool or Mean Pool for extracting the features. ICNN proposed in this paper uses the features of Max pool and Mean pool at different levels. The Max pool is used at the initial stage and the Min pool is used during the intermediate stages.

4   Comparison and Analysis

The performance of the proposed ICNN algorithm is presented in this section. The ICNN algorithm is evaluated against the traditional convolution algorithms that use either max pooling or average pooling. The ICNN algorithm uses both max pooling and average Pooling. This helps the algorithm to converge quickly with a greater speed and accuracy than the other competent algorithms. For evaluation a set of 4000 sample human images downloaded from google and also from INRIAPerson - http://pascal. inrialpes.fr/data/human/ are used for training the proposed algorithm. The algorithm is compared against the traditional CNN and RCNN. Residual Convolution Neural Network algorithm is used by Microsoft for image classification.

From Table 4, it is evident that out the 284 human images, the traditional CNN is able classify only 183 images correctly. The RCNN which is primarily used in space craft navigation can identify only 175 images correctly, out of 284. The ICNN proposed in this paper is able to identify 267 human images correctly out of 284. The evaluation is based on the accuracy, validation and the loss.

Table 4. Comparison of Performance of CNN, RCNN and ICNN

Classification algorithm	Number of sample images used for testing	Number of images identified correctly
CNN	284	183
RCNN	284	175
ICNN (Proposed)	284	267

The training accuracy depends on the number of stages required for the Improved Convolution Neural Network to get trained for image identification. From Fig. 7, it is noted that the proposed ICNN has improved accuracy with minimum number of stages (Fig. 8).

The loss function detects the number of losses that occurs during the image classification process. The major objective of loss function is to prevent the algorithm from being locked at the local minima. From Fig. 9 it is clearly evident that the ICNN does not struck at the local minima. This proves that the ICNN outperforms other two algorithms. The percentage of classification accuracy in different algorithms is represented in Fig. 10.

Accuracy

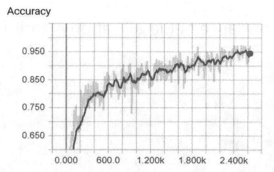

Fig. 7. Graph showing the training accuracy in ICNN

Accuracy/Validation

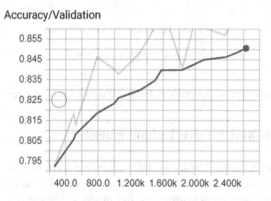

Fig. 8. Graph showing the training accuracy validation for ICNN

Adam/Loss/raw

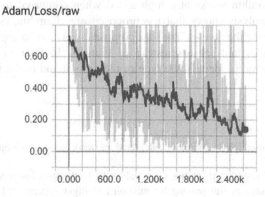

Fig. 9. Graph showing the effect of loss function in ICNN

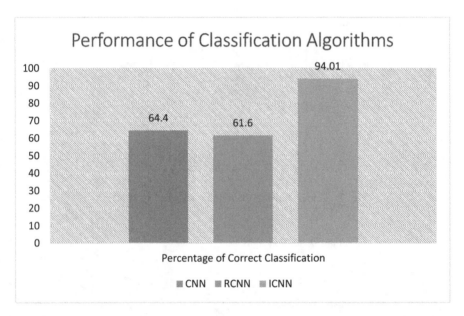

Fig. 10. Percentage of Accuracy in Image classification by different algorithms

5   Conclusions and Future Enhancements

The Cloud Enabled Country Border Intruder Detection System using ICNN proposed in this work is a novel idea that combines the features of WINS and ICNN. The WINS used in this work is made up of sensors and IoT devices. The use of fire base cloud and MQTT enables fast and secure way for exchanging messages. The use of ICNN that combines the features of both max pool and average pool, helps to detect the intruders instantly. The algorithm works at a high speed when compared to other competent algorithms. The analysis shows that the proposed algorithm shows around 25% of increase in classifying the images. The idea proposed by us can be applied to secure our border using devices and sensors instead of humans. In future we would like to improve the algorithm to identify the type of the arms found in the intruders.

References

1. Pottie, G.J., Kaiser, W.J.: Wireless integrated network sensors. Commun. ACM **43**(5), 51–58 (2000)
2. Vardhan, S., Wilczynski, M., Pottie, G.J., Kaiser, W.J.: Wireless integrated network sensors (WINS): distributed in situ sensing for mission and flight systems. IEEE (2000). 0-7803-5846-51001$10.00 0
3. Shire, I., Payal, K., Katkhede, K.N., Mande, K.P.: A review on security of border using WINS. Int. J. Eng. Sci. Res. Technol. **6**(8), 109–111 (2017). Accessed 5 Aug 2017

4. Yim, J., Ju, J., Jung, H., Kim, J.: Image classification using convolutional neural networks with multi-stage feature. In: Kim, J.-H., Yang, W., Jo, J., Sincak, P., Myung, H. (eds.) Robot Intelligence Technology and Applications 3. AISC, vol. 345, pp. 587–594. Springer, Cham (2015). https://doi.org/10.1007/978-3-319-16841-8_52

5. Tripathi, G., Singh, K., Vishwakarma, D.K.: Convolutional neural networks for crowd behaviour analysis: a survey. Vis Comput. Int. J. Comput. Graph. (2018). https://doi.org/10.1007/s00371-018-1499-5

6. Anuse, A., Vyas, V.: A novel training algorithm for convolutional neural network. J. Complex Intell. Syst. **2**(3), 221–234 (2016)

7. Cao, Y.S., Niu, X., Dou, Y.: Region-based convolutional neural networks for object detection in very high resolution remote sensing images. In: International Conference on Natural Computation, Changsha, China (2016)

8. Ding, P., Zhang, Y., Deng, W.J., Jia, P., Kuijper, A.: A light and faster regional convolutional neural network for object detection in optical remote sensing images. ISPRS J. Photogramm. Remote. Sens. **141**, 208–2018 (2018)

9. Srivastava, N., Shree, U., Chauhan, N.R., Tiwari, D.K.: Firebase cloud messaging. Int. J. Innov. Res. Sci. Eng. Technol. **6**(9), 1–8 (2017)

10. Soni, D., Makwana, A.: A Survey on MQTT: A Protocol of Internet of Things (IoT), Conference Paper (2017)

11. Firebase Cloud Messaging. https://firebase.google.com/docs/cloud-messaging. Accessed 27 July 2018

12. Message Queuing Telemetry Transport. https://www.pubnub.com/blog/what-is-mqtt-use-cases. Accessed 27 July 2018

13. Kendall, J.: Controlling cross-border traffic without sacrificing security (2018)

14. del Rio, J.S., Moctezuma, D., Conde, C., de Diego, I.M., Cabello, E.: Automated border control e-gates and facial recognition systems. J. Comput. Secur. **62**, 49–72 (2016)

15. Nunamaker Jr., J.F., Burgoon, J.: Sensors for Intelligent Monitoring of Human Interactions (2017)

16. Alom, M.Z., Hasan, M., Yakopcic, C., Taha, T.M., Asari, V.K.: Improved Inception-Residual Convolutional Neural Network for Object Recognition (2017)

17. He, K., Zhang, X., Ren, S., Sun, J.: Deep residual learning for image recognition. Microsoft Research (2015)

Temporal and Stochastic Modelling of Attacker Behaviour

Rahul Rade[(✉)], Soham Deshmukh, Ruturaj Nene, Amey S. Wadekar,
and Ajay Unny

Veermata Jijabai Technological Institute, Mumbai, India
{rsrade_b15, ssdeshmukh_b15}@el.vjti.ac.in,
ruturajnene97@gmail.com, wadekaramey1997@gmail.com,
ajayunny2013@gmail.com

Abstract. Cyber Threat Analysis is one of the emerging focus of information security. Its main functions include identifying the potential threats and predicting the nature of an attacker. Understanding the behaviour of an attacker remains one of the most important aspect of threat analysis, much work has been focused on the detection of concrete network attacks using Intrusion Detection System to raise an alert which subsequently requires human attention. However, we think inspecting the behavioural aspect of an attacker is more intuitive in order to take necessary security measures. In this paper, we propose a novel approach to analyse the behaviour of an attacker in cowrie honeypot. First, we introduce the concept of Honeypot and then model the data using semi-supervised Markov Chains and Hidden Markov Models. We evaluate the suggested methods on a dataset consisting of over a million simulated attacks on a cowrie honeypot system. Along with proposed stochastic models, we also explore the use of Long Short-Term Memory (LSTM) based model for attack sequence modelling. The LSTM based model was found to be better for modelling of long attack sequences as compared to Markov models due to their inability to capture long term dependencies. The results of these models are used to analyse different attack propagation and interaction patterns in the system and predict attacker's next action. These patterns can be used for a better understanding of the existing or evolving attacks and may also aid security experts to comprehend the mindset of an attacker.

Keywords: Cyber security · Threat intelligence · Cowrie honeypot
Markov chain · Hidden Markov Models · Attacker behavioral analysis
Sequence modelling using LSTM

1 Introduction

"In warfare, information is power. The better you understand your enemy, the more able you are to defeat him" [1]. In order to defeat attackers, security professionals need to learn and understand the attacker's methods, tactics, motivation and tools. Although information technology (IT) systems provide a rich set of functionalities, they are also susceptible to a more challenging set of threats each unique to the attacker. The need to protect and secure critical infrastructure has led to recent developments in threat

© Springer Nature Singapore Pte Ltd. 2019
L. Akoglu et al. (Eds.): ICIIT 2018, CCIS 941, pp. 30–45, 2019.
https://doi.org/10.1007/978-981-13-3582-2_3

analytics. Different threat analysis methods have been used to identify the elements of a threat against critical infrastructure assets and also analyse its behavioural patterns.

In order to execute useful response action, the deployed models should be able to detect future or ongoing intrusion attacks to the system. The current research literature mainly focuses on developing sophisticated Intrusion Response System (IRS) to satisfy this requirement [2]. The most important part determining the effectiveness of an IRS is its prediction model which helps the system to take accurate countermeasures before the intrusion happens. Thus, an accurate prediction of attacker's action will help the system to isolate and contain the attack without excessively affecting the whole network. Moreover the response should be targeted specifically to the attacker.

In this work, we focus on accurately modelling the attacker behaviour and predicting his next action. We propose a model based on Markov Chain and HMM for behaviour analysis on cowrie honeypot data for short sequences. Later, we focus on Long Short Term Memory (LSTM) architecture to analyse long sequence data. We model the behaviour of an attacker using the mentioned models and use these models to predict future behaviour of an attacker. This can aid the organization in taking the necessary preventive measures through human intervention or automated system.

The key contributions of this work are:

- Successfully modelling the behaviour of an attacker who tries to penetrate a file based Cowrie honeypot system.
- Providing an estimate of the next action or step the attacker might take and thus, enable the honeypot to take appropriate actions to slow down an attacker or prevent his actions.
- Detailed analysis of the honeypot logs providing insight on attacker behaviour and trends.

1.1 Related Work

A lot of efforts are put in understanding the user interaction with the system and finding underlying pattern. Researchers have proposed generative and discriminative models to achieve this goal. Previously one of the approach is used to understand the user browsing behaviour modelled by Markov models [3]. Also, a lot of previous work has been focused on classifying web sessions into "attack" and "normal" classes with the help of Intrusion Detection System (IDS) wherein these web sessions were usually modelled using supervised machine learning algorithms that were used to distinguish sessions [4–6]. Similarly, previous works have also focused on the analysis of polymorphic worms which can be considered complementary work to our suggested method [7, 8]. In some of these methods, signature models are used to identify the worms [7]. While these methods are necessary they do not provide better understanding of the nature of existing and future threats. Having the ability to predict attacker actions is of paramount importance for taking preventive rather than damage control based actions. Thus, in our work, we use Markov chain and Hidden Markov Model, along with LSTMs, to model different types of attacker behaviour that can be observed in a common file based honeypot system.

2   Cowrie Honeypot Dataset and Logging

Honeypot is a decoy system [9] that can simulate one or many vulnerable hosts, with the intent of tricking the attacker with an easy target. The honeypot does not have any other role to fulfil and therefore all connection attempts are deemed suspicious. In this work, we use the logs generated by cowrie honeypot for evaluating our approach.

Cowrie [19] is a medium interaction SSH and Telnet honeypot designed to log brute force attacks and the shell interaction performed by the attacker. Using Cowrie honeypot, user activity is logged in JSON files with distinct focus on terminal inputs and messages. The default JSON logging format employed is shown below:

```
{
        "eventid": "cowrie.client.version",
        "macCS": ["hmac-sha1", "hmac-sha1-96", "hmac-md5", "hmac-md5-96",
                "hmac-ripemd160", "hmac-ripemd160@openssh.com"],
        "timestamp": "2017-07-03T18:30:08.235671Z",
        "session": "577076dfb79e",
        "kexAlgs": ["diffie-hellman-group14-sha1", "diffie-hellman-group-
                exchange-sha1", "diffie-hellman-group1-sha1"],
        "keyAlgs": ["ssh-rsa", "ssh-dss"],
        "message": "Remote SSH version: SSH-2.0-PUTTY",
        "system": "HoneyPotSSHTransport,523,116.31.116.16",
        "isError": 0,
        "src_ip": "116.31.116.16",
        "version": "SSH-2.0-PUTTY",
        "dst_ip": "10.10.0.13",
        "compCS": ["none"],
        "sensor": "DC-NIC-Mumbai",
        "encCS": ["aes128-ctr", "aes192-ctr", "aes256-ctr", "aes256-cbc",
                "rijndael-cbc@lysator.liu.se", "aes192-cbc", "aes128-cbc",
                "blowfish-cbc", "arcfour128","arcfour", "cast128-cbc","3des-cbc"]
}
```

eventid is a tag given to the activity of the user, in our case this implies the hacker is checking the version of cowrie honeypot

timestamp: the exact time of login or the action being performed in

message: command prompt logs

session: current session id

src_port: the source port

system: honeypot system which is being intruded

isError: did the given action performed lead to any error

src_ip: source IP of the attacker

dst_ip: destination IP

The JSON file is parsed into a csv file for further processing and exploration. The dataset was also explored for outliers and abnormalities which can subsequent affect the performance for further modelling. The following points were considered:

- The occurrence of particular 'event id'
- The source IPs which are hitting the honeypot the most
- Most common passwords used while logging in the honeypot
- The frequency of hit on different honeypot systems
- The most common commands used in terminal.

As shown in Fig. 1, cowrie.command.success is the most common action being recorded indicates the actions of hacker are being completely executed. This implies that the malicious action of hackers are recorded in logs without fail. Cowrie.command. input is then the second most common action being taken. This includes the actions of deleting, downloading and injecting files into the file system.

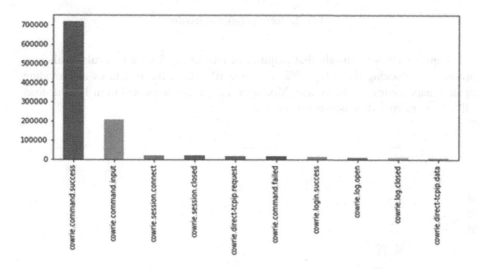

Fig. 1. Count of most occurring events in attack sessions

On further analysis of cowrie.command.input, the most common type of attack performed by hacker is a delete operation followed by input. On further analysis one can exactly determine if hackers are after some particular file or folder, and are they interested in perturbing it with virus or directly deletion (Fig. 2).

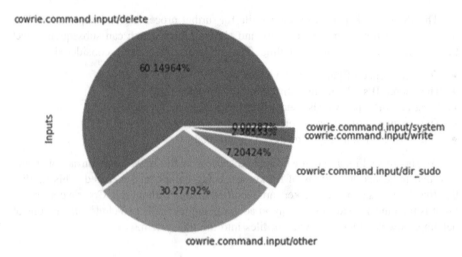

Fig. 2. Input events distribution

Figure 3 gives an insight that majority of attacks are from a particular malicious source IP, blocking those top 20% of source IP will be the first tasks to providing preliminary security to the system. Moreover, the attacks originated from 1924 distinct IP addresses spreading across 90 countries.

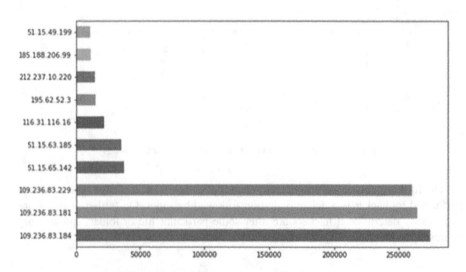

Fig. 3. Most common IPs involved in attacks

The most attacked system in the honeypot are shown in Fig. 4 which provides insight on which is a higher target for hackers and the priority of securing systems.

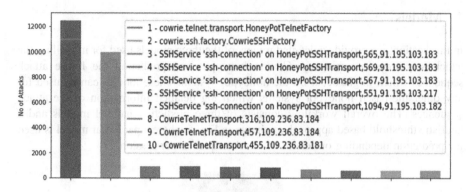

1 - cowrie.telnet.transport.HoneyPotTelnetFactory
2 - cowrie.ssh.factory.CowrieSSHFactory
3 - SSHService 'ssh-connection' on HoneyPotSSHTransport,565,91.195.103.183
4 - SSHService 'ssh-connection' on HoneyPotSSHTransport,569,91.195.103.183
5 - SSHService 'ssh-connection' on HoneyPotSSHTransport,567,91.195.103.183
6 - SSHService 'ssh-connection' on HoneyPotSSHTransport,551,91.195.103.217
7 - SSHService 'ssh-connection' on HoneyPotSSHTransport,1094,91.195.103.182
8 - CowrieTelnetTransport,316,109.236.83.184
9 - CowrieTelnetTransport,457,109.236.83.184
10 - CowrieTelnetTransport,455,109.236.83.181

Fig. 4. Most attacked honeypot systems

Honeypot Logs Dataset Processing for Modelling

The logged cowrie activity was modelled further into 14 states with focus on input commands. The input command was then divided into 5 commands specifically write, system, sudo, delete and other with prominently emphases the malicious behaviour of attacker. The total 19 states are given below along with their corresponding IDs.

{
    0: 'cowrie.client.size',
    1: 'cowrie.client.version',
    2: 'cowrie.command.failed',
    3: 'cowrie.command.input/delete',
    4: 'cowrie.command.input/dir_sudo',
    5: 'cowrie.command.input/other',
    6: 'cowrie.command.input/system',
    7: 'cowrie.command.input/write',
    8: 'cowrie.command.success',
    9: 'cowrie.direct-tcpip.data',
    10: 'cowrie.direct-tcpip.request',
    11: 'cowrie.log.closed',
    12: 'cowrie.log.open',
    13: 'cowrie.login.failed',
    14: 'cowrie.login.success',
    15: 'cowrie.session.closed',
    16: 'cowrie.session.connect',
    17: 'cowrie.session.file_download',
    18: 'cowrie.session.input'   }

The data is grouped by session id for considering each sequence where each session id corresponds to the sequence of actions taken by hacker. The assumption made here is different session id are independent of individual attacker characteristics and hence dividing depending on session ID rather than source IP wouldn't affect the modelling and prediction by a large factor.

3 Models

In this section, we provide a detailed review of the three models used for modelling the attack sequences occurring on Cowrie honeypot and predicting the future attack sequences which might occur on the honeypot system. These models can be used for any other file-based honeypot systems which log detailed information of the attack sequences. The overall workflow consists of three prediction parallel models and a heuristic threshold based approach as shown in Fig. 5 where a particular model is used for prediction depending on length of attack sequence per session.

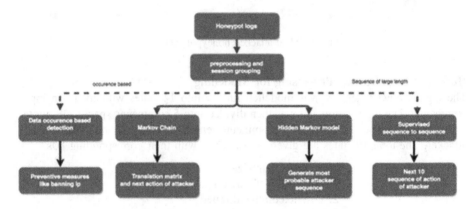

Fig. 5. Proposed workflow

3.1 Markov Chain

A Markov chain [12] is a stochastic process that sequentially moves from one state to another in the state space. A Markov chain [13] is defined by a state space $X = \{x_i: i = 0, 1, \ldots\}$ and the probabilities of transition between the states in the state space. In case of a Markov chain with a finite number of distinct states, these transition probabilities can be represented by a matrix that is always non-negative and where the sum of elements in each row equals one. Such a transition matrix is called one-step transition matrix. The main distinction of a Markov chain is the existence of memory which can help in efficiently modelling systems which follow a series of states, where the next state depends on the current state of the system.

In context of analysing honeypot attack sequences, the main idea is to train a probabilistic model which efficiently models the intruder attack sequences on Cowrie honeypot. A first-order Markov chain having states $X = \{x_i: i = 0, 1, \ldots, 18\}$ is being used for the purpose, where each state represents the ID of the event logged by the honeypot in response to the action taken by the attacker. According to a first order Markov chain, the probability of a state X at instant t is dependent only on the state at instant $t - 1$ i.e.,

$$P\left(X_t = x \mid X_{t-1}, X_{t-2}, X_{t-3}....X_1, X_0\right) = P\left(X_t \mid X_{t-1}\right) \tag{1}$$

where, X_t is the state of the model at instant t and $x \in X$.

We are interested in finding the probabilities $P\left(X_i \mid X_j\right)$ that if X_j is the current step which the attacker takes then the next step would be Xi. Training a Markov model involves calculating these transition probabilities using the number of transitions from state i to state j in the training sequences. Thus, the transition probability from state X_j to state X_i is calculated as

$$P\left(X_i \mid X_j\right) = \text{no of transitions from state j to state i/total no of transitions} \tag{2}$$
from state j to all states

The resulting Markov chain reveals the sequence of steps taken by an intruder while attacking a Cowrie honeypot system. Moreover, it can be used to predict the next step which the attacker might take depending on the current step of the attacker.

3.2 Hidden Markov Model

The Hidden Markov Model (HMM) have gained popular over past few years. HMM [10, 14] is a statistical Markov model with unobserved state. In HMM the states are not visible, but the output depends on states that are visible which makes it fast and useful in Cyber security systems.

An HMM is characterized by following:

1. N, the number of hidden states. All hidden states are interconnected which means any state can be reached from other states. The individual states are denoted by $S = \{S_1, S_2, ... S_N\}$ and state at time t is given by q_t.
2. M, the total number of observable states. These are the actual observations in application per state. In this application the observable states would be the different actions taken by the user. There are total of 19 observable states. The states are denoted by $V = \{v_1, v_2, ...v_M\}$
3. The state transition probability as $A = \{a_{ij}\}$ where

$$a_{ij} = P\left[q_{t+1} = S_i \mid q_t = S_i\right] \tag{3}$$

4. The observation symbol probability distribution in state, $B = \{b_j(k)\}$, where

$$b_j(k) = p\left[v_k \text{ at } t \mid q_t = S_j\right] \qquad 1 \leq j \leq N \tag{4}$$

5. The initial state distribution $\pi = \{\pi i\}$ where

$$\pi = P[q_1 = Si] \qquad 1 \leq i \leq N \tag{5}$$

The parameterized HMM can be used to predict the probability of observed sequence or to generate the given sequence. The HMM is completely specified by specification of two model parameters (N and M), specification of observation symbols, and the specification of the three initial probability measures A, B, and π. The compact notation is given by

$$\lambda = (A, B, \pi) \tag{6}$$

Baum Welch Algorithm

The Baum Welch algorithm [11] is used to find out the parameters of the model using Maximum Likelihood Estimation. It uses forward-backward algorithm to make the estimations when the states are given. The goal is to determine $\lambda = \{a_{ij}, b_j(k), \pi_i\}$ given sequence of state as $O_1, O_2, \ldots O_d$.

The algorithm (BW) used for selecting the parameter values belongs to a family of algorithms called Expectation Maximization (EM) algorithms. The likelihoods are used to re-estimate the parameters, repeatedly until a local maximum is reached.

For a given sequence, Od, probability of transiting from state i to j at time t is

$$P\left(q_t^d = i, q_{t+1}^d = j \mid O^d, \lambda\right) = \frac{P\left(q_t^d = i, q_{t+1}^d = j, O^d\right)}{P\left(O^d\right)} = \frac{\alpha_t(i) a_{ij} b_j\left(O_{t+1}^d\right) \beta_{t+1}(j)}{P\left(O^d\right)} \tag{7}$$

From this we can estimate,

$$A_{ij} = \sum_d \frac{1}{P\left(O^d\right)} \sum_t \alpha(t, i) a_{ij} b_i\left(O_{t+1}^d\right) \beta(t+1, i) \tag{8}$$

The probability of Od can be estimated using current parameter values using the Forward algorithm. Similarly,

$$E_i(\sigma) = \sum_d \frac{1}{P\left(O^d\right)} \sum_{X\{t \mid O_t^d = \sigma\}} \alpha(t, i) \beta(t, i) \tag{9}$$

From A_{ij} and $E_i(\sigma)$ we re-estimate the parameters of the model.

The Hidden Markov model used in for modelling was trained using Expectation Maximization algorithm. The model has been given 19 observable states and 19 hidden states. The initialization is chosen at random as to match it to the dataset. We implemented variable length Markov model to accommodate variable length of attack sessions and its higher order characteristic. We employed Baum-Welch algorithm to estimate the parameters of the model and trained the model on all the observed attack sessions. The sub-functions of the algorithm namely, forward and backward algorithms were implemented in Cython to minimize the training time. The Baum-welch algorithm function returns the probability estimation sequence containing the measures for every iteration.

The detailed algorithm of Baum-Welch algorithm and the detailed explanation can be found in [11].

3.3 Generative Sequence Modelling Using LSTM

Markov chain or hidden Markov model fall short on modelling sequences of large number of actions. Long Short-Term Memory (LSTM) or Long Short-term memory is a modified version of recurrent neural network introduced by Hochreiter and Sch-midhuber [16]. LSTM unit is composed of a cell, an input gate, an output gate and a forget gate. It forms a chain like structure where each cell can be assumed to be a time frame [16]. LSTM solves the vanishing gradient problem commonly faced in RNN which allows it to remember long sequences because LSTM architecture allows disabling of writing to a cell by turning off the gate, thus preventing any changes to the contents of the cell over many cycles [15, 18] (Fig. 6).

$$f_t = \sigma_g\left(W_f x_t + U_f h_{t-1} + b_f\right) \tag{10}$$

$$i_t = \sigma_g(W_i x_t + U_i h_{t-1} + b_i) \tag{11}$$

$$o_t = \sigma_g(W_o x_t + U_o h_{t-1} + b_o) \tag{12}$$

$$c_t = f_t \times c_{t-1} + i_t \times \sigma_g(W_o x_t + U_o h_{t-1} + b_o) \tag{13}$$

$$h_t = o_t \times \sigma_h(c_t) \tag{14}$$

where,

x_t: input vector to the LSTM unit
f_t: forget gate's activation vector
i_t: input gate's activation vector
o_t: output gate's activation vector
h_t: output vector of the LSTM unit
c_t: cell state vector

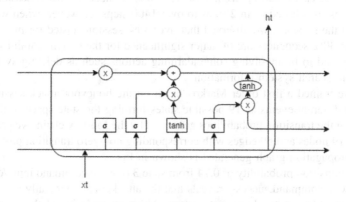

Fig. 6. LSTM cell

These long-term dependencies can greatly influence the predictions of future action as most proficient hacker lurk in the system doing benign actions for a long time. This make it essential to be able to capture this sequence of actions, as a false negative in this case would greatly affect the system.

Formatting and modelling data in a supervised manner was essential for working of LSTM. The processing of data was done as follows:

1. Each attack sequence consisting of various actions was padded to a fixed length of hundred and ten future actions were considered as output. A sliding window approach was used for maximum utilization of data.
2. Input sequences of action less than hundred and output sequences of action less than ten were padded with 0.
3. The output was then normalized between one and zero and 20% of data was reserved as testing data.

The architecture of model consisted of two LSTM layers, where the output of first LSTM is fed to second LSTM. In the second layer of LSTM, the outputs of previous cells are ignored and only the output of last cell is taken. The output activation function is tanh and returns a output of dimension (1, 10). The problem was framed as regression instead of classification. Mean Squared Error is used as loss function and the model is trained of 1000 epochs with a batch size of 256.

For evaluating the model, the predicted model score was floored and then the predicted value and truth value were used to compute accuracy. Hyperparameter like sequence length played a important role in the ability of LSTM to predict future sequence length.

4   Results

In this section, we present the results from applying the suggested methods on a massive data set of honeypot attack sequences. The dataset was extracted from the logs which the Cowrie honeypot generated from April 2017 to July 2017. Based on these logs, we generated 22,499 distinct attack sessions involving the series of steps taken by a particular source ip starting from cowrie.session.connect to cowrie.session.closed. The attack sessions lasted from 2 steps to over 1400 steps. However, when we further investigated the sessions, we observed that over 30% sessions lasted for more than 20 steps. These 30% sequences are of major significance for the system administrators as they were found to be involving some alarming actions such as deleting system files and trying to search system information.

First, we trained a first order Markov chain on the honeypot attack sessions considering 19 different events as 19 possible states forming the state space of the chain. Investigating the transition probabilities generated by the Markov chain, we produced a graph with 19 nodes and 69 edges with corresponding non-zero transition probabilities. The state propagation graph generated is shown in Fig. 7.

A high transition probability of 0.74 from state 3 (cowrie.command.input/delete) to state 8 (cowrie.command.success) reveals that the attackers are generally successful in deleting some files on the honeypot system. Although majority of the states in the

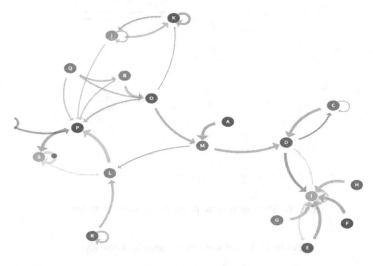

Fig. 7. Attack Propagation Graph generated using Markov chain

attack sessions involve success and input states, only 2.12% were found to be involving downloading files on the honeypot system while 1% states involved deleting files on the system. In addition to these statistics, the Markov chain also provides an insight of the further actions the attacker is about to take.

The results of Hidden Markov Model were similar to that obtained from Markov Chain. But testing on real time logs of attacker actions for 1 month (which we did not use for training) showed a prediction accuracy of 77% as compared to Markov chain which had 72% accuracy. The most probable attack sequence of length 14 generated by HMM is presented in Fig. 8 along with the corresponding sequence of hidden states. Figure 9 shows the convergence of the Hidden Markov Model being trained using expectation maximization algorithm.

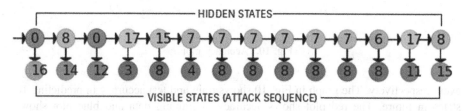

Fig. 8. Most probable sequence of length 14 generated by HMM

The accuracies of these two stochastic approaches along with that of LSTM based generative model are presented in Table 1.

It can be seen that LSTM outperforms Markov chain and Hidden Markov Model on large sequences of actions. Given previous 100 actions, LSTM sequence model predicts the future 10 states in time with an accuracy of 86 to 74% for 1st event to 10th

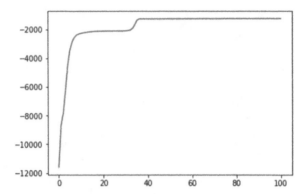

Fig. 9. The logarithmic probabilities every iteration

Table 1. Model used vs Percentage Accuracy

Model	Accuracy
Markov chain	72
HMM	77
LSTM	86

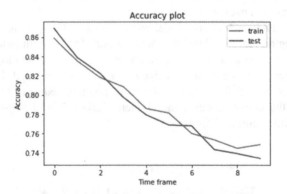

Fig. 10. Accuracy plot

event respectively. The graph in Fig. 10 shows train and test accuracy in predicting 10 states in future. The red plot shows accuracy of training data and blue plot shows accuracy of test data. Here timeframe refers the states prediction in future time.

An added advantage of LSTM based model is that it can predict 10 future states with a much greater accuracy; even the accuracy for 10th future state predicted by LSTM is comparable to the accuracy for 1st future state predicted by Markov chain. For example, the future 10 states predicted by LSTM based model for an input sequence [16, 14, 12, 3, 3, 8, 3, 2, 3, 2, 3, 8, 5, 8, 8, 8, 8, 8, 5, 8, 8, 5] are [8, 8, 6, 8, 8, 6, 8, 8, 5, 7], whereas the actual 10 stated are [8, 8, 5, 8, 8, 5, 8, 8, 5, 8].

The predictions of these models along with the actual attack sequences are shown below.

Actual Attack Seq. [16, 14, 12, 3, 8, 3, 2, 3, 2, 3, 8, 3, 8, 3, 8, 8, 8, 3, 8, 8, 8, 11, 15]
Markov Chain   [16, 14, 12, 3, 8, 8, 8, 3, 8, 3, 8, 8, 8, 8, 8, 8, 8, 8, 8, 8, 8, 8, 8, 15]
HMM   [16, 14, 12, 3, 8, 3, 8, 3, 8, 3, 8, 3, 8, 3, 8, 8, 8, 3, 8, 8, 8, 3, 15]
LSTM   [16, 9, 11, 3, 8, 3, 2, 3, 2, 3, 8, 3, 8, 3, 8, 8, 8, 3, 8, 8, 8, 3, 12]

Actual Attack Seq. [16, 14, 12, 3, 8, 3, 2, 3, 2, 3, 8, 3, 8, 3, 8, 8, 8, 3, 8, 8, 8, 4, 8, 8,
                    8, 8, 8, 8, 4, 8, 8, 8, 8, 8, 8, 4, 8, 8, 8, 8, 8, 8, 4, 8, 8, 8, 8, 8, 8, 4,
                    8, 8, 8, 8, 8, 8, 4, 8, 8, 8, 8, 8, 8, 4, 8, 8, 8, 8, 8, 8, 4, 8, 8, 8, 8, 8,
                    8, 4, 8, 8, 8, 8, 8, 8, 4, 8, 8, 8, 8, 8, 8, 4, 8, 8, 8, 8, 8, 4, 8, 8, 8, 8,
                    8, 8, 4, 8, 8, 8, 8, 8, 4, 8, 8, 8, 8, 8, 8, 4, 8, 8, 8, 8, 8, 8, 3, 8, 4,
                    8, 8, 8, 4, 8, 8, 8, 4, 8, 8, 8, 4, 8, 8, 8, 4, 8, 8, 8, 4, 8, 8, 8, 4, 8, 8,
                    8, 4, 8, 8, 8, 4, 8, 8, 8, 4, 8, 8, 8, 4, 8, 8, 8, 4, 8, 8, 8, 4, 8, 8, 8, 4,
                    8, 8, 8, 5, 8, 7, 8, 8, 8, 8, 8, 8, 3, 8, 8, 3, 8, 11, 15]
Markov Chain   [16, 14, 12, 3, 8, 8, 8, 3, 8, 3, 8, 8, 8, 8, 8, 8, 8, 8, 8, 8, 8, 8, 8, 8, 8,
                    8, 8, 8, 8, 8, 8, 8, 8, 8, 8, 8, 8, 8, 8, 8, 8, 8, 8, 8, 8, 8, 8, 8, 8, 8, 8,
                    8, 8, 8, 8, 8, 8, 8, 8, 8, 8, 8, 8, 8, 8, 8, 8, 8, 8, 8, 8, 8, 8, 8, 8, 8, 8,
                    8, 8, 8, 8, 8, 8, 8, 8, 8, 8, 8, 8, 8, 8, 8, 8, 8, 8, 8, 8, 8, 8, 8, 8, 8, 8,
                    8, 8, 8, 8, 8, 8, 8, 8, 8, 8, 8, 8, 8, 8, 8, 8, 8, 8, 8, 8, 8, 8, 8, 8, 8, 8,
                    8, 8, 8, 8, 8, 8, 8, 8, 8, 8, 8, 8, 8, 8, 8, 8, 8, 8, 8, 8, 8, 8, 8, 8, 8, 8,
                    8, 8, 8, 8, 8, 8, 8, 8, 8, 8, 8, 8, 8, 8, 8, 8, 8, 8, 8, 8, 8, 8, 8, 8, 8, 8,
                    8, 8, 8, 8, 8, 8, 8, 8, 8, 8, 8, 8, 8, 8, 8, 8, 8, 15]
HMM   [16, 3, 3, 3, 8, 3, 8, 3, 8, 3, 8, 3, 8, 3, 8, 3, 8, 4, 8, 3, 8, 4, 8, 8, 8,
                    8, 8, 8, 4, 8, 8, 8, 8, 8, 4, 8, 8, 8, 8, 8, 8, 4, 8, 8, 8, 8, 8, 8, 4, 8,
                    8, 8, 8, 8, 4, 8, 8, 8, 8, 8, 4, 8, 8, 8, 8, 8, 8, 4, 8, 8, 8, 8, 8, 8, 8,
                    4, 8, 8, 8, 8, 8, 4, 8, 8, 8, 8, 8, 8, 4, 8, 8, 8, 8, 8, 8, 8, 8, 8,
                    8, 4, 8, 8, 8, 8, 8, 8, 4, 8, 8, 8, 8, 8, 8, 4, 8, 8, 8, 8, 8, 8, 4, 8, 3, 8,
                    8, 8, 8, 8, 8, 8, 4, 8, 8, 8, 8, 8, 8, 8, 4, 8, 8, 8, 8, 8, 8, 4, 8, 8, 8,
                    8, 8, 8, 8, 4, 8, 8, 8, 8, 8, 8, 8, 8, 8, 4, 8, 8, 8, 8, 8,
                    8, 8, 4, 8, 8, 8, 8, 8, 4, 8, 3, 8, 3, 8, 8, 3, 3]
LSTM   [16, 9, 11, 4, 8, 3, 2, 3, 2, 3, 8, 3, 8, 4, 8, 8, 8, 3, 8, 8, 8, 4, 8, 8,
                    8, 8, 8, 8, 4, 8, 8, 8, 8, 8, 8, 4, 8, 8, 8, 8, 8, 8, 4, 8, 8, 8, 8, 8, 8, 4,
                    8, 8, 8, 8, 8, 8, 4, 8, 8, 8, 8, 8, 8, 4, 8, 8, 8, 8, 8, 8, 4, 8, 8, 8, 8, 8,
                    8, 4, 8, 8, 8, 8, 8, 8, 4, 8, 8, 8, 8, 8, 8, 4, 8, 8, 8, 8, 8, 4, 8, 8, 8, 8,
                    8, 8, 4, 8, 8, 8, 8, 8, 4, 8, 8, 8, 8, 8, 8, 4, 8, 8, 8, 8, 8, 8, 3, 8, 5,
                    8, 8, 8, 4, 8, 8, 8, 4, 8, 8, 8, 4, 8, 8, 8, 4, 8, 8, 8, 4, 8, 8, 8, 4, 8, 8,
                    8, 4, 8, 8, 8, 4, 8, 8, 8, 4, 8, 8, 8, 4, 8, 8, 8, 4, 8, 8, 8, 4, 8, 8, 8, 4,
                    8, 8, 8, 5, 8, 8, 8, 8, 8, 8, 8, 8, 4, 8, 8, 3, 8, 10, 11]

The effectiveness of LSTM for longer sequences is evident from the above examples. Additionally, it can be seen that HMM performs considerably better than Markov chain which gets biased towards high probability states. Moreover, the predictions of these models can provide a significant aid for taking preventive measures for improving the security of file based systems.

5 Conclusion and Future Work

In this work, we investigated the use of probabilistic models for modelling attack sequences occurring on Cowrie honeypots based on the logs generated by the honeypots. Further, we presented two novel approaches to effectively model the behaviour of an intruder, thereby providing an estimate of emerging threats posed by the professional attackers.

Using first order Markov chain along with variable length Hidden Markov Model, we were successfully able to model short to medium attack sequences. These trained models are capable of predicting next highly probable steps which the attackers might take along with an insight of the most common attacking trends and new and evolving techniques which attackers use. On the other hand, an LSTM model played a critical role in modelling long term dependencies present in sequences consisting of more than hundred actions and with a commendable accuracy can predict the future actions of attacker.

Analysing the sequences generated by the proposed models may shed light on new and evolving techniques which attackers take which intruding into a system. Thus, the system administrators may take more stringent steps to tackle these emerging threats. In addition, the developed algorithms can be extended further from file system logs and used to develop a retaliatory sequence of action with two end benefits. First to ban the attacker if he possesses an immediate threat to the system or any way could compromise the system. Second to stall the attacker in the honeypot long enough to generate his accurate logs. Furthermore, this works truly highlights the potential of development of a reinforcement-based algorithm wherein the agent is a system which can take preventive or retaliatory actions and environment consists of attackers and his action. Massive number of interactive logs will play a pivotal role in learning an appropriate policy based on traditional bellman equation.

Acknowledgement. We acknowledge the support of Centre of Excellence (CoE) in Complex and Nonlinear Dynamical Systems (CNDS), VJTI and Larsen & Toubro Infotech (LTI) under their 1-Step CSR initiative.

References

1. Schneier, B.: Honeypots and the Honeynet Project (2001). http://www.cs.rochester.edu/~brown/Crypto/news/3.txt. Accessed 26 July 2018
2. Cheng, B.C., Liao, G.T., Huang, C.C., Yu, M.T.: A novel probabilistic matching algorithm for multi-stage attack forecasts. IEEE J. Sel. Areas Commun. **29**(7), 1438–1448 (2011)
3. Shukla, D., Singhai, R.: Analysis of users web browsing behavior using Markov chain model. Int. J. **2**, 824–830 (2010)
4. Norouzian, M.R., Merati, S.: Classifying attacks in a network intrusion detection system based on artificial neural networks - IEEE Conference Publication. Paper presented at the 13th International Conference on Advanced Communication Technology (ICACT 2011), Seoul, South Korea, 13–16 February 2011 (2011)

5. Masduki, B.W., Ramli, K., Saputra, F.A., Sugiarto, D.: Study on implementation of machine learning methods combination for improving attacks detection accuracy on Intrusion Detection System (IDS). Paper presented at the 2015 International Conference on Quality in Research (QiR), Lombok, Indonesia, 10–13 August 2015 (2016)
6. Kim, K., Aminanto, M.E.: Deep learning in intrusion detection perspective: overview and further challenges. Paper presented at the 2017 International Workshop on Big Data and Information Security (IWBIS), Jakarta, Indonesia, 23–24 September 2017 (2018)
7. Kolesnikov, O., Lee, W.: Advanced Polymorphic Worms: Evading IDS by Blending in with Normal Traffic (2005): CC Technical report; GIT-CC-05-09, Georgia Institute of Technology. http://hdl.handle.net/1853/6485. Accessed 26 July 2018
8. Koganti, V.S., Galla, L.K., Nuthalapati, N.: Internet worms and its detection. Paper presented at the 2016 International Conference on Control, Instrumentation, Communication and Computational Technologies (ICCICCT), Kumaracoil, India, 16–17 December 2016 (2018)
9. Hong, J., Hua, Y.: IOP Conference Series: Materials Science and Engineering, vol. 322 052033 (2018). http://iopscience.iop.org/article/10.1088/1757-899X/322/5/052033/pdf. Accessed 26 July 2018
10. Rebiner, L.R.: A tutorial on hidden Markov models and selected applications in speech recognition. In: Proceedings of the IEEE (1989)
11. Hoberman, R., Durand, D.: HMM Lecture Notes (2006). http://www.cs.cmu.edu/~durand/03-711/2006/Lectures/hmm-bw.pdf. Accessed 26 July 2018
12. Grinstead, C.M., Snell, J.L.: Introduction to probability. American Mathematical Society (2012)
13. Chan, K.C., Lenard, C.T., Mills, T.M.: An Introduction to Markov Chains (2012). https://doi.org/10.13140/2.1.1833.8248
14. Rabiner, L.R., Juang, B.-H.: An introduction to hidden Markov models. ASSP Mag. 3(1), 4–16 (1986)
15. Cho, K., et al.: Learning phrase representations using RNN encoder-decoder for statistical machine translation. In: Proceedings of the Empirical Methods in Natural Language Processing (EMNLP 2014) (2014, to appear)
16. Graves, A.: Generating sequences with recurrent neural networks (2013). arXiv:1308.0850 [cs.NE]
17. Bengio, Y., Frasconi, P., Simard, P.: The Problem of Learning Long-Term Dependencies in Recurrent Networks, pp. 1183–1195. IEEE Press, San Francisco (1993)
18. Hochreiter, S., Schmidhuber, J.: Long short-term memory. Neural Comput. 9(8), 1735–1780 (1997)
19. Official repository for the Cowrie SSH and Telnet Honeypot effort. https://github.com/micheloosterhof/cowrie. Accessed 26 July 2018
20. Pascanu, R., Mikolov, T., Bengio, Y.: On the difficulty of training Recurrent Neural Networks (2013). arXiv:1709.03082v7 [cs.NE] 10 Mar 2018

Automatic License Plate Recognition Using Deep Learning

Bhavin Dhedhi[(✉)], Prathamesh Datar, Anuj Chiplunkar, Kashish Jain,
Amrith Rangarajan, and Jayshree Kundargi

K. J. Somaiya College of Engineering, Department of Electronics
and Telecommunication Engineering, Mumbai, India
{bhavin.dhedhi,prathamesh.datar,anuj.c,kashish.j,
amrith.r,jmkundargi}@somaiya.edu

Abstract. Automatic License Plate Recognition (ALPR) has been a topic of research for many years now due to its real-life application but hasn't been any significant breakthrough due to limitations in image processing algorithms to satisfy all the real-life scenarios such an illumination, moving cars, background etc. This paper presents a robust and efficient ALPR system using a combination of the 'You only Look Once' (YOLO) neural network architecture and standard Convolutional Neural Network (CNN). In total 3 stages of YOLO and 1 stage of CNN has been used in the proposed system. The last stage of YOLO and CNN have been specifically designed to perform detection (segmentation) and recognition of characters, respectively. We have built our own dataset of 604 car images in natural settings with different lighting conditions and viewing angles for the YOLO stages. In addition, a computer-generated dataset of 42237 characters has been used to train CNN. The resulting system has been tested on 50 random test images not part of training or validation datasets. The validation accuracies of all 4 stages exceed 90% whereas, the overall final accuracy on 50 test images comes to 82% with some fault tolerance. The use of deep learning instead of Image Processing also enabled to detect skewed license plates. The accuracy of stages 1 and 2 of YOLO were 100% on both validation and test sets.

Keywords: ALPR · Deep learning · Image processing · YOLO
CNN · Dataset · Accuracy

1 Introduction

Automatic License Plate Recognition (ALPR) or Number Plate Recognition (NPR) are methods which utilize optical character recognition (OCR) to identify vehicle license plates in an efficient manner without the need for major human resource investment and have steadily become more important in the recent years as urban societies grapple with the issue of traffic congestion and insecure corporate and residential car parks. Automatic License Plate Recognition Systems use the concept of optical character recognition to read the characters on a vehicle license plate. In other words, ALPR takes the image of a vehicle as the input and outputs the characters present on its license plate. The rapid development in image processing techniques has also made it possible to detect and identify license plates at a fast rate.

© Springer Nature Singapore Pte Ltd. 2019
L. Akoglu et al. (Eds.): ICIIT 2018, CCIS 941, pp. 46–58, 2019.
https://doi.org/10.1007/978-981-13-3582-2_4

Deep Learning architectures are artificial intelligence architectures that are used to imitate the way the human brain processes data by forming patterns to help in decision making. Deep learning utilizes a hierarchical level of artificial neural networks to facilitate the process of learning from unsupervised, unstructured and unlabeled data. The artificial neural networks are built to process data like the human brain, having nodes called as 'neurons', connected to each other to create a web-like structure. Unlike traditional programs which analyze data linearly, the hierarchical nature of deep learning neural networks allows machines to approach processing of large amounts of data in a nonlinear fashion.

As mentioned earlier, the ALPR system's task is to recognize the license plate from the image or video stream and recognize the characters in the license plate by looking them up in a database. The requirement for such an ALPR system is high accuracy when reading the license plates and reasonably fast processing time.

A typical ALPR system can be split into four major stage:

1. Car detection – detect the car in the captured image.
2. License plate detection – detect the plate in the captured image.
3. Character segmentation/detection – extract the alphanumeric characters from the plate.
4. Character recognition – recognize each individual character.

Most of the ALPR systems use proprietary software like MATLAB on host computers to accomplish each of these steps. This paper presents an alternative approach to automatically recognize license plates using open source soft wares like Open Computer Vision Library (OpenCV) and Python and Neural Network architectures- YOLO (You Only Look Once!) and CNN (Convolutional Neural Network).

Deep learning techniques do not use hand-engineered features, but automatically select the features themselves. They are designed to learn low-level representations of the underlying data by modifying filter coefficients during each training epoch. The strongest deep learning methods involve You Only Look Once (YOLO) and Convolutional Neural Networks (CNNs).

YOLO is a state-of-the-art neural network architecture used for real-time object detection. This network first divides an input image into a number of regions and predicts bounding boxes and their probabilities for each region. These boxes formed are then assigned weights by the predicted probabilities. A big advantage of YOLO is that it considers the entire image at the time of testing, so its predictions take into consideration the global context of the input. Unlike other networks which require thousands of network evaluations, YOLO makes its makes predictions with a single network evaluation.

CNNs are hierarchical neural networks based on sparse connections and weight sharing giving them an immense representational capacity and high learning potential. The biggest challenges of CNNs are their high computational cost and demand for large amounts of training samples.

Identification of vehicles is used for many different operations. It can be used by government agencies to find cars that are involved in crime, look up if annual fees are paid or identify persons who violate the traffic rules. France, United States, Singapore,

Japan, Germany, Italy, and U.K are all countries that have successfully applied ALPR in their traffic management. The functionalities of the proposed ALPR system can be used for Automatic Toll Collection System and can be further enhanced by incorporating security features to alert the authorities when any unauthorized number plate is detected using buzzer alarm system. ALPR can also be used conveniently used in parking lots to monitor the cars moving in and out of the parking lot.

This paper is composed of several sections: Sect. 2 gives us the previous related work in this field i.e. Literature Survey. Section 3 gives the information about the training and testing datasets. Section 4 discusses the proposed system. Section 5 discusses the training parameters used for the networks. Experimentation and results are given in the Sect. 6. A complete conclusion is drawn in the Sect. 7. Section 8 highlights the future aspects of the paper.

2   Literature Survey

In this section, we briefly analyze the recent works in the context of ALPR. For Image processing-based works, please refer to [1–11]. The ML approaches are discussed with respect to each ALPR stage in the pipeline. This subsection concludes with final remarks.

LP Detection
Many researchers have addressed the LP detection stage with object detection CNNs. In [12] a single CNN was used in a cascaded manner to detect both front view of the car and its License Plates, achieving high accuracy. In [13] Support Vector Machines (SVM) and Region-based CNN (RCNN) was used for License Plate detection. In [14] customized CNNs were built exclusively for LP detection. In [15] a CNN trained with character from general text was trained to perform LP detection. The results achieved were highly accurate than previous approaches.

License Plate Segmentation
In [12] a customized CNN was used to segment and recognize the characters within a cropped License Plate. The CNN segmented 99% of the characters correctly, outperforming the baseline by a large margin. In [16] high accuracy License plate segmentation and recognition was achieved using Hidden Markov Models (HMM).

Character Recognition
Deep CNNs were used for character recognition in [17, 18]. Deep Learning makes character recognition less sensitive to spatial transformations. In [17] random CNNs were used to extract features for character recognition, achieving a significantly better performance than using image pixels or learning the filters weights with backpropagation. In [15] a Recurrent Neural Network (RNN) is employed to label the sequential data, recognizing the whole LP without the character-level segmentation.

Miscellaneous
IN [19] an end to end ALPR pipeline system using a sequence of YOLO networks was used. This system was geared towards Brazilian plates. In [21] a commercial ALPR system using a sequence of Deep CNN's was presented. As this is a commercial system, little information is given about the used CNNs. Li et al. [20] propose a unified

CNN that can locate LPs and recognize them simultaneously in a single forward pass. In addition, the model size is highly decreased by sharing many of its convolutional features.

Final Remarks

Many papers only address part of the ALPR pipeline (e.g., LP detection) or perform their experiments on datasets that do not represent real-world scenarios, making it difficult to accurately evaluate the presented methods. In this sense, we employ the YOLO object detection CNNs in the first three stages to create a robust and efficient end-to-end ALPR system.

3 Dataset

A. Vehicle Detection

The dataset for vehicle detection contains in total 604 images of Car. These images consist of a combination of both stationary and moving cars with varied distance and angle between the car and the camera. The images were taken over varied backgrounds like Parking Lots, Roads and traffic and different lighting conditions to account for real life scenarios. The cameras used were: One Plus 2 and Lenovo Z2 Plus. Although, the cameras have different autofocus, optical image stabilization and focal length, all the images are available in Joint Photographic experts Group Format (JPEG) with a resolution of 4160×3120 pixels.

B. License Plate Detection

Same set of 604 images were used for the LP detection stage. These images covered LP's of different physical sizes and word length. The images accounted for different orientations, skewness and location of license plate in the image.

C. Character Segmentation

The dataset consists of 4181 images of License Plate. Out of these, 581 images belonged to our dataset and 3600 images of LP's belonged to the Brazilian UFPR-ALPR dataset.

D. Character Detection

In this stage we used a combination of computer generated fonts and cropped characters from our dataset. In Total 42273 images were used.

Apart from the above, 50 images were used for the testing of the ALPR pipeline. Every image has the following annotations available in a XML file: The vehicle's position, position of the LP, as well as the position of its characters.

In India, each State uses a set of two particular starting letters for its LPs which results in a specific range of LP's for a given state. With a huge population of vehicles, the nomenclature of LP's is very complex. Also the number of characters in the dataset my may vary leading to more difficulty. Additionally, the non-standardization of LP's add to the complexity of building an ideal dataset. In Maharashtra (where the dataset was collected), LPs range from MH-01-AA-0001 to MH-50-ZZ-9999. The word length varies between 9 and 10 characters depending upon the LP. Therefore, the letters M and H have many more examples than the others. We also managed to include cars with LP of others states to make our dataset more general purpose (Figs. 1, 2 and 3).

Fig. 1. Sample images of dataset. These images show the diversity in the position of car, LP position, backgrounds, and lighting conditions. LPs blurred for privacy.

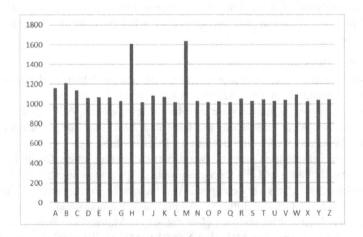

Fig. 2. Letters Distribution our dataset

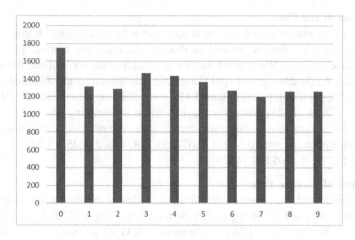

Fig. 3. Number Distribution in our dataset

4 Proposed System

The pipeline that we used for ALPR is to break the entire process into four separate steps. The first step detects the presence of a car in an image. The second step detects the license plate, the third step segments characters on the plate and the last step recognizes each individual character. We used 3 separate YOLO networks to perform the first three steps and then used a modified CNN to recognize characters in the last step. This entire process is visualized in Fig. 4.

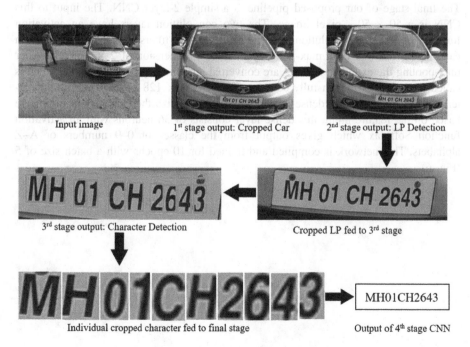

Fig. 4. Proposed System pipeline

A. Detection of the Car

The first stage of our proposed system is to detect a car in the given frame. The idea behind first detecting the car instead of directly detecting the number plate was to minimize false positive. We wanted for our network to only look inside the car for a number plate. To do this we first split the dataset into training and validation sets with 80/20 split. So, the training set contained 484 and validation set contained 120 images. We trained YOLO v2 [25] (As shown in Fig. 5) on this training set.

For better accuracy, we used our own anchor boxes which we created using K-Means algorithm and are: [0.57273, 0.677385, 1.87446, 2.06253, 3.33843, 5.47434, 7.88282, 3.52778, 9.77052, 9.16828].

B. Detection of License Plate

All the setting for this stage were kept the same as that of the first stage except for the anchor boxes. We used following anchor boxes for this stage: [1.28, 0.43, 1.96, 0.63, 2.52, 0.84, 3.16, 1.03, 4.00, 1.36]. We trained YOLO v2 on 604 images with corresponding license plate bounding boxes. At the time of testing our proposed system, we only used the cropped car which was outputted by the first stage.

C. Detection of Characters

Third YOLO v2 is used for detection and extraction of characters from the image of the number plate. We trained this network on 4181 images of license plate (581 plates from our dataset + 3600 from [19]) with anchor boxes: [0.05, 0.16, 0.08, 0.21, 0.11, 0.29, 0.17, 0.37, 0.25, 0.51]. These detected characters were then passed to final stage CNN for recognition.

D. Final stage CNN

The final stage of our proposed pipeline is a simple 2-layer CNN. The input to this CNN is a 50×50 - pixel image. The two convolution layers have an activation function of ReLU. Convolution is carried out with 64 filters of dimension 5×5 in each layer, followed by a max-pool layer with the dimension of 2×2 pixels. After max-pooling the extracted features are converted into an array of 1×40000, which is called flattening. A hidden fully connected layer of 1×128 nodes is used after flattening. Finally, an output dense layer is used to recognize the character in the image based on the feature input, this dense layer consists of 36 neurons with an activation function Softmax which gives output from the classes of 0–9 numbers or A–Z alphabets. This network is compiled and trained for 10 epochs with a batch size of 5 (Fig. 6).

No	Layer	Filter	Size	Input	Output
0	conv	32	3 x 3/1	416 x 416 x 3	416 x 416 x 32
1	max		2 x 2/2	416 x 416 x 32	208 x 208 x 32
2	conv	64	3 x 3/1	208 x 208 x 32	208 x 208 x 64
3	max		2 x 2/2	208 x 208 x 64	104 x 104 x 64
4	conv	128	3 x 3/1	104 x 104 x 64	104 x 104 x 128
5	conv	64	1 x 1/1	104 x 104 x 128	104 x 104 x 64
6	conv	128	3 x 3/1	104 x 104 x 64	104 x 104 x 128
7	max		2 x 2/2	104 x 104 x 128	52 x 52 x 128
8	conv	256	3 x 3/1	52 x 52 x 128	52 x 52 x 256
9	conv	128	1 x 1/1	52 x 52 x 256	52 x 52 x 128
10	conv	256	3 x 3/1	52 x 52 x 128	52 x 52 x 256
11	max		2 x 2/2	52 x 52 x 256	26 x 26 x 256
12	conv	512	3 x 3/1	26 x 26 x 256	26 x 26 x 512
13	conv	256	1 x 1/1	26 x 26 x 512	26 x 26 x 256
14	conv	512	3 x 3/1	26 x 26 x 256	26 x 26 x 512
15	conv	256	1 x 1/1	26 x 26 x 512	26 x 26 x 256
16	conv	512	3 x 3/1	26 x 26 x 256	26 x 26 x 512
17	max		2 x 2/2	26 x 26 x 512	13 x 13 x 512
18	conv	1024	3 x 3/1	13 x 13 x 512	13 x 13 x 1024
19	conv	512	1 x 1/1	13 x 13 x 1024	13 x 13 x 512
20	conv	1024	3 x 3/1	13 x 13 x 512	13 x 13 x 1024
21	conv	512	1 x 1/1	13 x 13 x 1024	13 x 13 x 512
22	conv	1024	3 x 3/1	13 x 13 x 512	13 x 13 x 1024
23	conv	1024	3 x 3/1	13 x 13 x 1024	13 x 13 x 1024
24	conv	1024	3 x 3/1	13 x 13 x 1024	13 x 13 x 1024
25	skip_connection	64	1 x 1/1	13 x 13 x 1024	26 x 26 x 64
26	conv	1024	3 x 3/1	13 x 13 x 1024	13 x 13 x 1024
27	conv	30	1 x 1/1	13 x 13 x 1024	13 x 13 x 30

Fig. 5. YOLO v2 Architecture

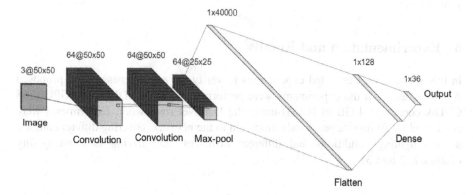

Fig. 6. Final stage CNN.

5 Training

We used Dell Inspiron 5577 laptop with 8 GB RAM and 4 GB GTX 1050 GPU, for the training of every stage.

First three stages:

The training process for all three YOLO stages is same. All the images are first converted into batches of 4 images each with the help of BatchGenerator. Learning rate is set as 1e−4. Now first we keep the warmup_epoch = 5 so that our network can familiarize with the dataset and there will be no premature stopping. After this step, we keep the nb_epoch to 50 and the algorithm starts training. We need to use both the warmup epochs as well as the nb_epochs otherwise there will be the premature stopping and we end up getting more than 200 boxes per image.

Other parameters like train_times and valid_times both are equal to 10 which indicates for how many times we'll go through the entire training and validation dataset in one epoch. In this way the training process takes place. we save the weights in the h5 format. We also use the Early Stopping so that it stops the training process when the validation loss does not decrease in the three consecutive epochs.

Apart from this the details of individual stage is given in the proposed system section.

CNN Training

Dataset for the CNN contains 42273 images for the classes 0–9 and A–Z which was developed by adding the computer-generated font with the extracted characters from the 581 Indian Plates of size 128×128. First the training and validation set is created by the 80–20 split. The input to the CNN network was 50×50 image. Two Convolution layers with the 'same' border mode and activation function of ReLu [23] was used. 64 filters of 5×5 kernel were used for training. For the maxpooling stage 2×2 kernel was used. Dense layer of 128 was applied after Flattening and a softmax classifier was used to get the probability of respective class. Dropout of 0.2 was set so that our model won't overfit the data and learn the generalised features of the characters. For the loss function we used categorical cross entropy and optimizer as adadelta [22]. We ran the entire network for 10 times and saved the weights in the h5 and yaml format.

6 Experimentation and Results

In this section, we conducted experiments to verify the effectiveness of the proposed ALPR system. All the experiments were performed on an NVIDIA 1050 GPU (768 CUDA cores and 4 GB of RAM) using the DarkNet framework. Experiments were conducted on 50 images previously unknown to the network, covering different angles, diverse lighting conditions and different distances, and having different quality (Tables 1, 2 and 3).

Table 1. No. of images used for training, validation and testing at each stage

Number of training & validation images for stages 1 & 2 of YOLOv2	604
Number of training & validation images for stage 3 of YOLOv2	4181
Number of training & validation images for stage 4 of CNN	42273
Number of testing images for the overall system	50

We report the results obtained by the proposed system. We perform evaluation by assessing individual subsections of the ALPR pipeline to verify their performance.

Table 2. Recognition rates of each stage of the proposed system on the testing dataset (in terms of images)

Stage	No. of correctly detected images	Total no. of images	Accuracy
YOLO 1- Car detection	50	50	100
YOLO 2- License plate detection	50	50	100
YOLO3- Character detection	41	50	82
CNN- Character recognition (1 incorrect character tolerance)	41	50	82
CNN- Character recognition (0 fault tolerance)	24	50	48

Table 3. Recognition rates of last two stages of the proposed system on the testing dataset (in terms of characters)

Stage	No. of correctly detected characters	Total no. of characters	Accuracy
YOLO3- Character detection	480	495	96.96
CNN- Character recognition	451	480	93.76

Analysing the above results, we realized that the accuracy of the overall system suffered when the plate was unclean (i.e., if it was dusty or if it had caught rust). This created a problem during the binarization of the plate as some of the dusty white background would get binarized to black. Additionally, the nuts and bolts used to hold the plate in place also were detected as parts of a character by the third YOLO stage, and reduced the accuracy. In order to reduce the percentage of incorrectly detected characters, we set the threshold confidence level for the third stage of YOLO as 0.5. Finally, the non-standard nature of some of the characters on the license plate contributed towards the loss in accuracy of the last CNN stage as some of the character fonts were completely unknown to the network (Fig. 7).

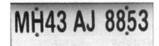

MH 43 AJ 8853

MH 03 CM 0661

MH 03 CM 7937

MH 03 AW 0384

MH 04 EH 775

7

Fig. 7. Examples of correctly and incorrectly recognized plates. Errors caused due to unclean plates and presence of nuts and bolts in segmented characters.

7   Conclusion

We have presented a robust ALPR system using the state to the art YOLO object detection and localization algorithm and a self-developed CNN network. We created our own dataset using 604 images of cars in the natural setting covering conditions like good lighting, bad lighting, moving cars, cars at a different distance from the camera, different orientations etc. This system provides improved security as we can detect plates only in the detected car region. At present, the bottleneck of ALPR systems is the character segmentation and recognition stages. To improve this, we used various data augmentation techniques such as Additive Gaussian Blur, Median Blur, Average Blur, etc. to improve the training performance. Seems fairly simple but is able to achieve outstanding results.

We realized that the Indian cars dataset was insufficient for efficient character segmentation stage, therefore, we added 3600 Brazilian Plate images to our 581 plate images and trained on it to get an exponentially improved accuracy for character segmentation.

Computer generated font was again insufficient to produce effective results. Therefore, we extracted characters from the 581 Indian Plates and added it to the computer-generated font to get a high increase in the accuracy of character recognition.

Our System was able to produce good results on the test set containing 50 images with the overall final accuracy of 82% with some fault tolerance. However, this result is still not practical for some real life ALPR applications. The main problem here is that although there is a legal format for License Plates in India, it is not followed judiciously.

The most difficult problem of ALPR applications is the correction of the skew. Through our proposed system we were also able to recognize license plates with skew with fairly high accuracy.

8 Future Work

The detection work is currently slow even though we are using real-time YOLO algorithm as it is taking a lot of time to load the weights in the system. The .h5 file format is not that efficient thus we'll try to implement the same using the weights format. We also intend to improve the character recognition accuracy by looking for more robust CNN architectures.

We wish to employ the Version 3 [24] of the YOLO which has an "objectness" score for each bounding box using logistic regression. Therefore, it can detect fairly large and small objects in the same image accurately.

We also plan to add temporal redundancy at the end of our system after character recognition. This will help reduce false positives thereby reducing the number of errors and improving accuracy. We also wish to improve our dataset. We wish to achieve this by adding more images of all kinds of vehicles not only limited to cars. We also wish to improve the diversity of our dataset by addition of images taken under different lighting conditions and angles. We also intend to add different license plates belonging to all the states in India. Additionally, we plan to explore the colour of the vehicle, vehicle's manufacturer and model in the ALPR pipeline as our new dataset provides such information. Finally, we intend to add live video capability in our ALPR pipeline.

Acknowledgment. We are elated to present this paper as it has extended our boundaries of knowledge and enhanced our capabilities and self-confidence. At the same time, we express our sincere thanks to everyone, who by their direct or indirect contribution have helped us make it possible. We would like to take this opportunity to express our gratitude towards our project guide Associate Prof. Jayshree Kundargi for her constant encouragement and guidance. We would also like to thank the Electronics and Telecommunication Department for providing us with valuable resources and aids as and when required.

References

1. Du, S., Ibrahim, M., Shehata, M., Badawy, W.: Automatic license plate recognition (alpr): A state-of-theart review. IEEE Trans. Circuits Syst. Video Technol. **23**(2), 311–325 (2013)
2. Zhou, W., Li, H., Lu, Y., Tian, Q.: Principal visual word discovery for automatic license plate detection. IEEE Trans. Image Process. **21**(9), 4269–4279 (2012)
3. Anagnostopoulos, C., Anagnostopoulos, I., Loumos, V., Kayafas, E.: A license plate-recognition algorithm for intelligent transportation system applications. IEEE Trans. Intell. Transp. Syst. **7**(3), 377–392 (2006). 16 Hui Li, Chunhua Shen
4. Bai, H., Liu, C.: A hybrid license plate extraction method based on edge statistics and morphology. In: Proceedings of the IEEE International Conference on Pattern Recognition, pp. 831–834 (2004)
5. Qiu, Y., Sun, M., Zhou, W.: License plate extraction based on vertical edge detection and mathematical morphology. In: Proceedings of the International Conference on Computer Intelligence Software Engineering, pp. 1–5 (2009)
6. Zheng, D., Zhao, Y., Wang, J.: An efficient method of license plate location. Pattern Recogn. Lett. **26**(15), 2431–2438 (2005)

7. Tan, J., Abu-Bakar, S., Mokji, M.: License plate localization based on edge-geometrical features using morphological approach. In: Proceedings of the IEEE International Conference on Image Processing, pp. 4549–4553 (2013)
8. Lalimia, M.A., Ghofrania, S., McLernonb, D.: A vehicle license plate detection method using region and edge based methods. Comput. Electron. Eng. **39**, 834845 (2013)
9. Chen, R., Luo, Y.: An improved license plate location method based on edge detection. In: Proceedings of the International Conference on Applications of a Physics Industrial Engine, p. 13501356 (2012)
10. Rasheed, S., Naeem, A., Ishaq, O.: Automated number plate recognition using hough lines and template matching. In: Proceedings of the World Congress on Engineering and Computer Science, pp. 199–203 (2012)
11. Deb, K., Jo, K.: Hsi color based vehicle license plate detection. In: Proceedings of the International Conference on Control Automation System, pp. 687–691 (2008)
12. Montazzolli, S., Jung, C.R.: Real-time brazilian license plate detection and recognition using deep convolutional neural networks. In: 2017 30th SIBGRAPI Conference on Graphics, Patterns and Images (SIBGRAPI), pp. 55–62, October 2017
13. Rafique, M.A., Pedrycz, W., Jeon, M.: Vehicle license plate detection using region-based convolutional neural networks. Soft Computing, June 2017. https://doi.org/10.1007/s00500-017-2696-2
14. Hsu, G.S., Ambikapathi, A., Chung, S.L., Su, C.P.: Robust license plate detection in the wild. In: 2017 14th IEEE International Conference on Advanced Video and Signal Based Surveillance (AVSS), pp. 1–6, August 2017
15. Li, H., Shen, C.: Reading car license plates using deep convolutional neural networks and lstms. CoRR, vol. abs/1601.05610, 2016
16. Bulan, O., Kozitsky, V., Ramesh, P., Shreve, M.: Segmentation and annotation-free license plate recognition with deep localization and failure identification. IEEE Trans. Intell. Transp. Syst. **18**(9), 2351–2363 (2017)
17. Menotti, D., Chiachia, G., Falcao, A.X., Neto, V.J.O.: Vehicle license plate recognition with random convolutional networks. In: 2014 27th SIBGRAPI Conference on Graphics, Patterns and Images, pp. 298–303, Aug 2014
18. Zhang, H., Jia, W., He, X., Wu, Q.: Learning-based license plate detection using global and local features. In: Proceedings of the IEEE International Conference on Pattern Recognition, pp. 1102–1105 (2006)
19. Rayson Laroca, Evair Severo, Luiz A. Zanlorensi, Luiz S. Oliveira, Gabriel Resende Goncalves, William Robson Schwartz, David Menotti, "Robust Real-Time Automatic License Plate Recognition Based on the YOLO Detector"
20. Li, H., Wang, P., Shen, C.: Towards end-to-end car license plates detection and recognition with deep neural networks. CoRR, vol. abs/1709.08828 (2017). http://arxiv.org/abs/1709.08828
21. Masood, S.Z., Shu, G., Dehghan, A., Ortiz, E.G.: License plate detection and recognition using deeply learned convolutional neural networks. CoRR, vol. abs/1703.07330 (2017). http://arxiv.org/abs/1703.07330
22. Zeiler, M.D.: ADADELTA: An Adaptive Learning Rate Method. arXiv:1212.5701 [cs.LG]
23. Xu, B., Wang, N., Chen, T., Li, M.: Empirical Evaluation of Rectified Activations in Convolutional Network. arXiv:1505.00853v2 [cs.LG], 27 November 2015
24. Redmon, J., Farhadi, A.: YOLOv3: An Incremental Improvement
25. Redmon, J., Farhadi,A.: YOLO9000: Better, Faster, Stronger. arXiv:1612.08242 [cs.CV] 25 DEC 2016

Data Management and Processing Technologies

Towards Reliable Storage for Cloud Systems with Selective Data Encryption and Splitting Strategy

Z. Asmathunnisa[1(✉)] and P. Yogesh[2]

[1] St. Anne's College of Engineering, Panruti, Tamilnadu, India
zasmath99@gmail.com
[2] College of Engineering, Anna University, Chennai, India
yogesh@annauniv.edu

Abstract. Nowadays, reliability and security have become serious issues in Information and Communication Technology (ICT) since more and more data and services are accessed from computational cloud. Since cloud is an open platform and accessed through public networks like the Internet, user's data become vulnerable for security attacks. Foul play of cloud operators to reach sensitive data of users is one of the serious issues that need wide consideration as it vividly reduces the adoptability of cloud computing. Many practical security challenges are arising due to the abundant volume of data. Time used up in data encryption heavily hinders the performance of cloud based systems since data transmission and data communication are slowed down due to the large amount of data to be encrypted and decrypted. To attain an adoptive performance altitude many applications reject data encryption. In this paper, we focus on privacy leakage issues, and promote security levels under predefined time and resource constraints. To this end, we propose a Selective Data Encryption and Splitting Strategy (SDE2S), a compact encrypting method to selectively encrypt data according to the privacy weight and execution time of data packages being sent. Also it randomly splits data into n parts and then performs XOR operations using different cipher keys in different cloud storage servers to protect users' private information from possible untrusted cloud operators. Here, we put forward an overview of the problem and describe the algorithms used in the proposed solution. At the end, we present our simulation results, which reveal the advantages and improvements of our scheme over other schemes.

Keywords: Reliability · Selective Data Encryption · Data security
Splitting · Untrusted cloud operators

1 Introduction

Voluminous data like big data and multimedia data have forced the users of ICT to migrate towards cloud computing since it is not possible for an individual to own the required resources and the cloud based systems are based on pay as you go model. Cloud computing has stretched into many fields and many new service deployment models have been provided to the public [1, 2], like mobile parallel computing [3–7] and

© Springer Nature Singapore Pte Ltd. 2019
L. Akoglu et al. (Eds.): ICIIT 2018, CCIS 941, pp. 61–74, 2019.
https://doi.org/10.1007/978-981-13-3582-2_5

distributed scalable data storage [8–13]. Recently, big data has developed to be broadly used in various business domains and many researches are exploring it [14, 15].

In spite of many advantages of using big data in cloud computing, protecting user's privacy data incur wide range of issues [16, 17]. One of the biggest concerns is caused due to transmitting huge volume of data unencrypted [18, 19]. Wireless communications avoid encrypted data transmission to gain fair performance level. As plain texts are at ease for attackers to seize information by using spoofing, jamming and monitoring [20], this may lead to privacy leakage issues.

In addition, the Mass Distributed Storage (MDS) has reached to its large scale and size in current years [21] due to its flexibility and adjustable computation. Much sought is secure distributed data storage [22–26], as it has many threats such as chances of malevolent attacks or misuse activities. The attacks can take place during data transmissions and data communication as well. The unforeseen operations can also take place at the cloud server side, which need to be controlled by some means to balance between the availability of services and the security level [27–31]. Hence, completely controlling the risks of distributed data in cloud systems is a difficult goal, which needs to be achieved in an effective manner [32–34].

This paper addresses these issues and proposes a method that attempts to selectively encrypt data in order to maximize the encryption rate and also deals with the problem of cloud operators misusing users' data. Hence this method paves the way to avoid cloud users' data discharge on the cloud side. The proposed model is called a Selective Data Encryption and Splitting Strategy (SDE2S), which is premeditated to protect data owners' privacy and aims to split data and stores the encrypted data to different cloud servers without causing excess overhead and latency.

SDE2S is using the following techniques: (i) Data packages are sorted according to privacy level and (ii) Selective Encryption Data Determination (SE2D) Algorithm that determines the data encryption alternatives based on timing and resource constraints. (iii) After identifying the data that can be encrypted, the data is split into n parts to store them in n different cloud servers after encryption. This is done to protect user's private information from un-trusted cloud operators. As a token, we have represented only two cloud servers in Fig. 1. Splitting data process is achieved by our proposed algorithm Split-Data-Based Encryption (SDBE) Algorithm.

The paper is structured as per the following order: Sect. 2 deals with various research works that have been carried out in the area of data security and privacy preservation in cloud based systems. Next, Sect. 3 provides an overview about the algorithms used in this paper. Mechanisms and the methodology used in this research work are revealed in Sect. 4. This section also illustrates the main algorithm designed for supporting SED2S approach. Section 5 explains the outcome of the experiment carried over and comparison with other related works. Finally, conclusions are drawn in Sect. 6.

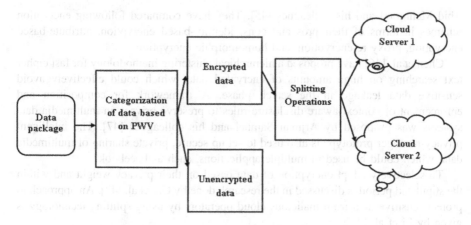

Fig. 1. High level architecture of SDE2S model

2 Related Work

Cloud is meant for information sharing between multiple users across different infrastructure. Communication among clients usually involves sensitive data via social media and infrastructure.

A methodology called SCLPV for cloud-based Cyber Physical Social Systems (CPSS) was proposed to protect from wicked auditors by Zhang et al. [35] by providing certificate less public verification to verify the integrity of outsourced data. To have a privacy preserving public auditing Wang et al. [36] developed a secure cloud system. These researches did not deal with the threats in the data transmission but focused on defining adversaries from the data storage side.

Numerous ways are there to create privacy issues in clouds. Among which unpredictable data is the root for privacy leakage creation since neither users nor service providers identify them because of the following reasons: (i) It is difficult to recognize the collected data as low trustworthy data and (ii) failing to provide any identification information regarding the data such as difficulty to generate threat alerts etc. creates privacy issues [37]. Next, to identify trustworthy data with large data size, the data analytic technique is considered a potential solution which has been discussed in the work [38].

Vulnerability detection is also a vital aspect of preventing privacy leakage [39, 40]. Exploitation of vulnerabilities induced by Graphical User Interface (GUI) elements has been detected by Mulliner et al. [41]. Secure searching system is incorporated in big data by accomplishing privacy policy compliance checking mechanism [42, 43]. Usage of an efficient secure networking system has been suggested to reduce the rate of the threat amplifications in [44]. Wireless networks pave many chances to intrude and to steal data while transmitting thus providing chances for stealing privacy by the untrusted parties.

Comprehensive survey of the privacy-protection methods in big data, including data generation, data storage, and data processing phases are offered by

Abid Mehmood and his colleagues [45]. They have compared following encryption schemes in terms of their pros and cons: identity-based encryption, attribute-based encryption, proxy re-encryption, and homomorphic encryption.

Chen et al. [46] have proposed a hierarchical clustering methodology for fast cipher text searching on huge amounts of encrypted data, which could effectively avoid sensitive data leakage in the search phase. A framework for composition and enforcement of context-aware disclosure rules to preserve privacy in multimedia data systems was proposed by Arjman Samuel and his colleagues [47]. The intelligent privacy-manager prototype is also used to set up secure, private sharing of multimedia data, which could be used in multiple applications, such as Facebook.

To selectively adopt encryption of data based on their privacy weight and within the stipulated period is discussed in the research done by Gai et al. [48]. An approach to protect sensitive data from malicious cloud operators by using splitting technology is given by Li et al. [49].

3   Our Solution

We introduce Selective Data Encryption and Splitting Strategy (SDE2S) model to promise privacy and security in cloud based systems. The working of this model can be summarized as follows: it first sorts out the data packages into multiple groups based on the level of Privacy Weight Value (PWV), and then decides whether a data package needs to be encrypted or not by employing Selective Encryption and Determination (SE2D) algorithm.

Selective Encryption Data Determination (SE2D) algorithm works on the parameters related to PWV and the operation time of every data package to choose the order of the encryption. By adopting another algorithm Split-Data-Based Encryption (SDBE), data packages to be encrypted are first randomly split into n separate components, and then transmitted to different cloud storage servers to avoid users' personal information being directly revealed to a single cloud operator who may become malicious in course of time.

When users want to retrieve back data from cloud servers, the proposed system retrieves encrypted data from different cloud servers and executes decryption operations.

The major features of our model are as follows:

1. The model considers the constraints of time delay and resources to assign the maximum total privacy weights from a set of variables containing the amount of data package types, the privacy weight for each data package, and the time required to deal with the data package with encryption and without encryption

2. The model employs a split-data-based encryption method to thwart malicious cloud operators from abusing the private data of the users. This method performs the XOR operation with a generated cipher key.

4 Concepts of Proposed Work

4.1 Selective Data Encryption and Splitting Strategy (SDE2S) Model

Selective Data Encryption and Splitting Strategy (SDE2S) model is based on Utmost Data Packages below Timing Constraints (UDPTC) problem. The features of UDPTC are described as follows.

Inputs: types of data package {Pi}, the number of data in each data package type NPi, execution time for encrypting each single data TePi, execution time for not encrypting each single data TnPi, the privacy weight value for each data type PWPi.

Results: an output showing data need to be encrypted. Maximum total privacy weight value beneath a given timing constraint is gained from our proposed system.

The input to the algorithm comprises five variables. First variable is the input data that includes a group of packages categorized into various types, represented as a set {Pi}. Second argument is the number of data packages in every type Pi, represented as NPi. Third parameter is the execution mode and there are two kinds of execution modes: with Encryption Operation (OwE) and without Encryption Operation (OwNE). Fourth variable is the execution time of each data package Pi. If the data package is in OwE mode then execution time is TePi and if it belongs to OwNE the execution time is TnPi. Furthermore, we introduce the fifth variable Privacy Weight Value (PWV) of each data package type in order to decide the advantages that can be obtained by the system from encrypting data and we represent it as WPi. PWV is a decisive factor that shows the impact of applying various security levels. In our proposed model, the PWV is used as value to signify the privacy of each data package.

Data packages need to be encrypted and which need not to be encrypted may be the output of our SDE2S strategy. Suppose that the number of encrypted data packages for Pi is NePi. We have designed our scheme as a maximization problem and the scheme attempts to maximize the sum of PWV values as shown in Eq. (1). Encrypted data packages alone are considered in our model by neglecting unencrypted data packages as they do not need any privacy weights.

$$\text{Outcome} = \text{Max}\left(\sum(\text{NePi} \times \text{WPi})\right) = \text{Plan} \tag{1}$$

Total execution time should be smaller than the required timing constraint Tc is the condition to be satisfied. The length of Tc is decided by the condition as per the Eq. (2). The expression shows the least execution time of data operations, without including all encryptions.

$$\text{Tc} \geq (\text{NPi} \times \text{TnPi}) \tag{2}$$

Selective Data Encryption and Splitting Strategy (SDE2S) approach selects the data packages to be encrypted once it is implemented. The scheme organizes the encrypted data set as {Pk} and the non-encrypted data set as {Pj}. The total execution time for the data packages is given as in Eq. (3).

$$Ttotal = \sum(NePk \times TePk) + \sum(NnPj \times TnPj) \qquad (3)$$

Ttotal should satisfy the condition Tc $\geq$ Ttotal.

4.2   Working of Selective Data Encryption and Splitting Strategy (SDE2S)

The major phases of SDE2S model are shown in Fig. 2.

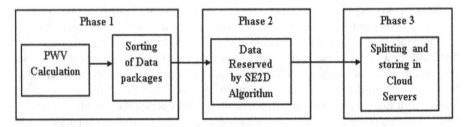

Fig. 2. Vital phases of Selective Data Encryption and Splitting Strategy (SDE2S) Model

Our Model constitutes three phases as below:

Phase I: Sorting of packages. This is the initial phase of the model. This phase sorts all data packages based on both their privacy value and execution time. Sorting of data packages are carried out based on two parameters: PWVs and encryption execution time. The variable utilized for sorting operations is represented by SPi and thus for every data package, SPi is obtained as shown in Eq. 4.

$$SPi = WPi/TePi \qquad (4)$$

The sorting results are entered in a table called S Table(ST).

Phase II: Data reserved. This phase is the decisive step meant for selecting data packages for encryption operations. Indication of protection efficiency is gained by using S Table(ST). The working principle of this phase is that the data package possessing higher value of SPi is given a higher-level reserving priority. If the given timing constraint is Tc, then the timing scope is [0, Ts], in which the value of Ts is computed as shown in Eq. (5).

$$Ts = Tc - \sum(NPi \times TnPi) \qquad (5)$$

Encryption time of a data package is represented as TePi. The operation will continue until two circumstances occur: Either all data packages are encrypted or the execution time TePi is not smaller than the remaining time. Remaining execution time is indicated as Tr, where Tr $\leq$ Ts. In our model, we calculate time Tr by taking into account both execution time with encryptions and execution time without encryption.

Once the data package is selected for encryption, the execution time without encryption should be added to Tr. Let us consider that the selected data packages for encryption are {Ps} and in that case the value of Tr is computed as in Eq. (6).

$$Tr = Ts - \sum (NPs \times TePs) + \sum (NPs \times TnPs) \tag{6}$$

The data reserved process ends when Tr is less than the execution time of any left data packages. Output of this phase gives out the plan of list and the order of reserved data packages to be encrypted.

Phase III: Splitting. Having obtained data reserved with maximizing total privacy weight values and security scores, we now look at split-data-based encryption strategy to utilize limited resources and time. This is achieved with the help of Split Data Based Encryption (SDBE) algorithm. The information may consist of data packages in form of string. Let X denotes the input data package reserved for encryption. X is first arbitrarily separated into two components namely A and B, and the relationship between A and B are represented as in Eq. (7).

$$B = X - A \tag{7}$$

Equation 7 is expected to satisfy the conditions $A \neq \varphi$ and $A \leq X$.

Now, A and B are sent out to cloud storage server 1 and cloud storage server 2, respectively, which helps us to put off sensitive data of the user from openly seep outing to cloud operators. The arbitrarily separated parts and generated cipher key values are always anonymous, which can greatly prevent malicious attackers and protect privacy. Ciphertext of A and B are generated by adopting RC2 algorithm as in Eqs. 8 and 9.

$$C = Ai\ XOR\ A \tag{8}$$

$$D = Bi\ XOR\ B \tag{9}$$

In these equations C and D represent the cipher texts of a data package with byte k, A and B represent the plaintext of a data package with byte k and Ai and Bi are the generated cipher keys with byte k. At last, store the generated cipher texts C and D in cloud storage server 1 and cloud storage server 2, respectively. This encryption method is used to resolve data package loss issues successfully.

Once the encryption phase is completed, we may want to retrieve back the original data from the cloud servers. It can be carried out easily by taking the encrypted data from cloud storage server 1 and cloud storage server 2 and perform XOR operations with the primary generated cipher keys. It could be expressed as in Eqs. 10 and 11.

$$E = Ai\ XOR\ C \tag{10}$$

$$F = Bi\ XOR\ D \tag{11}$$

Then, E is added to F to obtain the original data packages and there ends the privacy data decryption process.

4.3   Algorithms

We introduce the algorithms employed in our Selective Data Encryption and Splitting Strategy (SDE2S) Model. Selective Encryption Data Determination (SE2D) algorithm is intended to choose data packages those are capable to be encrypted and satisfying both timing constraints and resource availability. Key-in parameters of SE2D algorithm include Mapping Table (MT), Sort Table (ST), and Time Constraint (Tc). The result we get as output is the data encryption strategy plan P that expresses which data packages need to be encrypted. Algorithm 1 represents the pseudo code of SE2D algorithm.

Algorithm 1: Selective Encryption Data Determination (SE2D) algorithm

> **Key_In Values**: *ST, MT, Tc*
> **Output** : *Plan* (Data Encryption Strategy Plan)
> 1: Input *ST, MT,Tc*
> 2: Initialize *Plan* ←0
> 3: **do**
> 4:   Get *Di* having the maximum priority from ST
> 5:   **for** ∀ *Pi*, i=1 to *NPi* do
> 6:   if*Ts>TePi− TnPi* then
> 7:   Add one *Pi* to *Plan*
> 8:   *Ts← Ts− (TePi− TnPi)*
> 9:   else
> 10:   **Break**
> 11:   **end if**
> 12:   **end for**
> 13: **while** ST is not empty
> 14: Output *Plan*

Moreover, the outputs of SE2D algorithm should possess the highest value of total PWVs, Plan. The value of Plan can be derived from Eq. (1). The value of Total PWVs (TPWV) is $\sum$(NePi × WPi). Therefore, the expression of TPWV is represented as in Eq. (12).

$$TPWV = \sum (NePi \times WPi) \text{ for k bytes, where } i = 1 \text{ to k}$$
$$= NeP1WP1 + NeP2WP2 + \ldots + NekWPk \tag{12}$$

Algorithm 2 : Split-Data-Based Encryption (SDBE)Algorithm

Input: X, A
Output: C, D
1. Randomly choose one data package A to start;
2. **repeat**
3.   Employ RC2 algorithm to create cipher keys
4.   Generate cipher keys Ai ;
5.   **for** $\forall A$, **do**
6.   **if** $A \neq \omega$ and $A \leq X$ **then**
7. $B = X - A$;
8. $C = Ai$ XOR A
9. $D = Bi$ XOR B ;
10.   **else**
11.   **break**
12.   **end if**
13.   **end for**
14. **until** *the last data package encrypting in buffer;*

5 Experimental Results

In data reserving phase, we have evaluated our SED2S algorithm by setting up a simulation based on four different types of data packages (P1 to P4). Each type has a certain number of data packages. Each type of data package has different operation time. There are 2 modes: With encryption (M1) and without encryption (M2). Timing constraint Tc is 25 microseconds. The main aim is to get out the plan that can earn the highest total PWV by choosing a set of data packages for encryption. The operating principle is that a higher-level complexity of the data encryption will earn a higher PWV. The number of data packages for P1is 3 and the time consumed for a data package with encryption is 5-time units and the time consumed for a data package without encryption is 1-time unit and so on as shown in Table 1. We call this table as Mapping Table (MT).

Table 1. Mapping Table (MT)

Data Package Type	No. of data package in each type	M1		M2	
		TeDi	WDi	TnDi	WDi
P1	3	5	2.5	1	0
P2	4	3	2	0.5	0
P3	2	3	1.2	0.5	0
P4	2	4	1	1	0

Initially the values of SPi are calculated for the sorting purpose. We got results as SP1: 0.5, SP2: 0.67, SP3: 0.4, SP4: 0.25 and thus priority sequence is SP2, SP1, SP3, and SP4. Then Ts is calculated from (25 - (1 3 + 0.5 4 + 0.5 2 + 1 2)) as 17 units. As per Table 2, we create the following data encryption strategy: encrypt 4 P2, encrypt 1 P1, encrypt 1 P3, and do not encrypt P4. Thus the value of P is 11.7 = (4*2 + 1*2.5 + 1*1.2). Table 2 shows the data reserved for encryption.

Table 2. Table showing the data reserved for encryption

Tr	P2	P1	P3	P4	
17	0	0	0	0	No data is encrypted
14.5	1	0	0	0	1 P2 is encrypted
12	2	0	0	0	2 P2 are encrypted
9.5	3	0	0	0	3 P2 are encrypted
7	4	0	0	0	4 P2 are encrypted
3	4	1	0	0	4 P2 and 1 P1 are encrypted
0.5	4	1	1	0	4 P2, 1 P1 and 1 P3 are encrypted

Greedy algorithm can generate a strategy plan as follows: encrypt 3 P1, encrypt 1 P2, do not encrypt P3, and do not encrypt P4. Hence the value of P is 9.5. When comparing with greedy algorithm P value of our approach is 23.2% higher.

In the splitting phase, we evaluate the proposed split-data-based encryption method by simulating the operation time with different input data sizes. Two parameters were investigated, which had been data size and operational time in micro seconds. Figure 3 shows a comparison of the operation time of our data encryption and data retrieval methods with the Advanced Encryption Standard (AES) for data encryption, with input data sizes of 10, 50, and 200 Mbytes. The results show that the proposed data encryption and decryption method has the shortest operation time than AES.

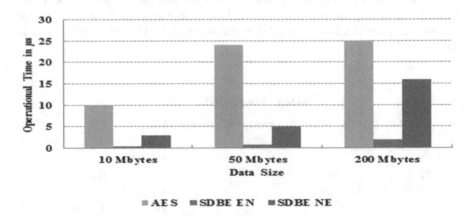

Fig. 3. The comparison of operation time with different data sizes

6 Conclusion

This paper focuses on the reliability of data storage in computational clouds by adopting selective encryption of data packages under time and resource constraints and splitting of data is intended to secure user's sensitive data reaching the curious cloud operators. To achieve this intention, we wished for a novel approach entitled as a Selective Data Encryption and Splitting Strategy (SDE2S) an efficient encrypting method to selectively encrypt the data resources according to the privacy weight and execution time of each data package. Also the scheme includes splitting data into n parts, generating cipher keys and performing XOR operations using generated cipher keys in different cloud storage servers to protect users' private information from the untrusted cloud operators. Our experimental evaluations have proved that the proposed scheme is more efficient than existing schemes.

References

1. Qiu, M., Zhong, M., Li, J., Gai, K., Zong, Z.: Phase-change memory optimization for green cloud with genetic algorithm. IEEE Trans. Comput. **64**(12), 3528–3540 (2015)
2. Gai, K., Li, S.: Towards cloud computing: a literature review on cloud computing and its development trends. In: 2012 Fourth International Conference on Multimedia Information Networking and Security, Nanjing, China, pp. 142–146 (2012)
3. Li, J., Ming, Z., Qiu, M., Quan, G., Qin, X., Chen, T.: Resource allocation robustness in multi-core embedded systems with inaccurate information. J. Syst. Archit. **57**(9), 840–849 (2011)
4. Chen, L., Duan, Y., Qiu, M., Xiong, J., Gai, K.: Adaptive resource allocation optimization in heterogeneous mobile cloud systems. In: The 2nd IEEE International Conference on Cyber Security and Cloud Computing, New York, USA, pp. 19–24. IEEE (2015)
5. Niu, J., Gao, Y., Qiu, M., Ming, Z.: Selecting proper wireless network interfaces for user experience enhancement with guaranteed probability. J. Parallel Distrib. Comput. **72**(12), 1565–1575 (2012)
6. Gai, K., Du, Z., Qiu, M., Zhao, H.: Efficiency-aware workload optimizations of heterogenous cloud computing for capacity planning in financial industry. In: The 2nd IEEE International Conference on Cyber Security and Cloud Computing, New York, USA, pp. 1–6. IEEE (2015)
7. Gai, K., Qiu, M., Zhao, H., Tao, L., Zong, Z.: Dynamic energy-aware cloudlet-based mobile cloud computing model for green computing. J. Network Comput. Appl. **59**, 46–54 (2015)
8. Gai, K., Qiu, M., Zhao, H.: Security-aware efficient mass distributed storage approach for cloud systems in big data. In: The 2nd IEEE International Conference on Big Data Security on Cloud, New York, USA, pp. 140–145 (2016)
9. Gai, K., Qiu, M., Tao, L., Zhu, Y.: Intrusion detection techniques for mobile cloud computing in heterogeneous 5G. In: Security and Communication Networks, pp. 1–10 (2015)
10. Wu, G., Zhang, H., Qiu, M., Ming, Z., Li, J., Qin, X.: A decentralized approach for mining event correlations in distributed system monitoring. J. Parallel Distrib. Comput. **73**(3), 330–340 (2013)

11. Yin, H., Gai, K.: An empirical study on preprocessing high dimensional class-imbalanced data for classification. In: The IEEE International Symposium on Big Data Security on Cloud, New York, USA, pp. 1314–1319 (2015)

12. Liang, H., Gai, K.: Internet based anti-counterfeiting pattern with using big data in china. In: The IEEE International Symposium on Big Data Security on Cloud, New York, USA, pp. 1387–1392. IEEE (2015)

13. Jean-Baptiste, H., Qiu, M., Gai, K., Tao, L.: Meta meta-analytics for risk forecast using big data meta-regression in financial industry. In: The 2nd IEEE International Conference on Cyber Security and Cloud Computing, New York, USA, pp. 272–277. IEEE (2015)

14. Li, Y., Gai, K., Ming, Z., Zhao, H., Qiu, M.: Intercrossed access control for secure financial services on multimedia big data in cloud systems. ACM Trans. Multimedia Comput. Commun. Appl. **PP**(99), 1 (2016)

15. Yin, H., Gai, K., Wang, Z.: A classification algorithm based on ensemble feature selections for imbalanced class dataset. In: The 2nd IEEE International Conference on High Performance and Smart Computing, New York, USA, pp. 245–249 (2016)

16. Li, Y., Dai, W., Ming, Z., Qiu, M.: Privacy protection for preventing data over-collection in smart city. IEEE Trans. Comput. **PP**, 1 (2015)

17. Gai, K., Qiu, M., Chen, L., Liu, M.: Electronic health record error prevention approach using ontology in big data. In: 17th IEEE International Conference on High Performance Computing and Communications, New York, USA, pp. 752–757 (2015)

18. Qiu, M., Gai, K., Thuraisingham, B., Tao, L., Zhao, H.: Proactive user-centric secure data scheme using attribute-based semantic access controls for mobile clouds in financial industry. Future Gener. Comput. Syst. **PP**, 1 (2016)

19. Gai, K., Qiu, M., Thuraisingham, B., Tao, L.: Proactive attribute-based secure data schema for mobile cloud in financial industry. In: The IEEE International Symposium on Big Data Security on Cloud, 17th IEEE International Conference on High Performance Computing and Communications, New York, USA, pp. 1332–1337 (2015)

20. Ma, L., Tao, L., Zhong, Y., Gai, K.: RuleSN: research and application of social network access control model. In: IEEE International Conference on Intelligent Data and Security, New York, USA, pp. 418–423 (2016)

21. Chang, F., et al.: Bigtable: a distributed storage system for structured data. ACM Trans. on Computer Systems **26**(2), 4 (2008)

22. Ateniese, G., Fu, K., Green, M., Hohenberger, S.: Improved proxy re-encryption schemes with applications to secure distributed storage. ACM Trans. Inf. Syst. Secur. **9**(1), 1–30 (2006)

23. He, X., Wang, C., Liu, T., Gai, K., Chen, D., Bai, L.: Research on campus mobile model based on periodic purpose for opportunistic network. In: 2015 IEEE 17th International Conference on High Performance Computing and Communications, New York, USA, pp. 782–785. IEEE (2015)

24. Gai, K., Steenkamp, A.: Feasibility of a Platform-as-a-Service implementation using cloud computing for a global service organization. In: Proceedings of the Conference for Information Systems Applied Research ISSN, vol. 2167, p. 1508 (2013)

25. Qiu, M., Sha, E.: Cost minimization while satisfying hard/soft timing constraints for heterogeneous embedded systems. ACM Trans. Des. Autom. Electron. Syst. **14**(2), 25 (2009)

26. Qiu, M., Cao, D., Su, H., Gai, K.: Data transfer minimization for financial derivative pricing using Monte Carlo simulation with GPU in 5G. Int. J. Commun. Syst. (2015)

27. Wu, G., Zhang, H., Qiu, M., Ming, Z., Li, J., Qin, X.: A decentralized approach for mining event correlations in distributed system monitoring. J. Parallel Distrib. Comput. **73**(3), 330–340 (2013)

28. Li, J., Qiu, M., Ming, Z., Quan, G., Qin, X., Gu, Z.: Online optimization for scheduling preemptable tasks on IaaS cloud systems. J. Parallel Distrib. Comput. **72**(5), 666–677 (2012)

29. Gai, K., Qiu, M., Tao, L., Zhu, Y.: Intrusion detection techniques for mobile cloud computing in heterogeneous 5G. Secur. Commun. Networks, 1–10 (2015)

30. Qiu, M., Gao, W., Chen, M., Niu, J., Zhang, L.: Energy efficient security algorithm for power grid wide area monitoring system. IEEE Trans. Smart Grid **2**(4), 715–723 (2011)

31. Zhao, H., Chen, M., Qiu, M., Gai, K., Liu, M.: A novel pre-cache schema for high performance Android system. Future Gener. Comput. Syst. (2015)

32. Li, Y., Chen, M., Dai, W., Qiu, M.: Energy optimization with dynamic task scheduling mobile cloud computing. IEEE Syst. J., 1–10, June 2015

33. Yu, X., Pei, T., Gai, K., Guo, L.: Analysis on urban collective call behavior to earthquake. In: The IEEE International Symposium on Big Data Security on Cloud, pp. 1302–1307, New York, USA. IEEE (2015)

34. Gai, K.: A review of leveraging private cloud computing in financial service institutions: value propositions and current performances. Int. J. Comput. Appl. **95**(3), 40–44 (2014)

35. Zhang, Y., Xu, C., Yu, S., Li, H., Zhang, X.: SCLPV: Secure certificateless public verification for cloud-based cyber-physical-social systems against malicious auditors. IEEE Trans. Comput. Soc. Syst. **2**(4), 159–170 (2015)

36. Wang, C., Chow, S., Wang, Q., Ren, K., Lou, W.: Privacy-preserving public auditing for secure cloud storage. IEEE Trans. Comput. **62**(2), 362–375 (2013)

37. Tang, L., et al.: A framework of mining trajectories from untrustworthy data in cyber-physical system. ACM Trans. Knowl. Discov. Data **9**(3), 16 (2015)

38. Schuster, F., Costa, M., Fournet, C., Gkantsidis, C., Peinado, M., Mainar Ruiz, G., Russinovich, M.: VC3: trustworthy data analytics in the cloud using SGX. In: IEEE Symposium on Security and Privacy, pp. 38–54, San Jose, CA, USA. IEEE (2015)

39. Maffei, M., Malavolta, G., Reinert, M., Schroder, D.: Privacy and access control for outsourced personal records. In: IEEE Symposium on Security and Privacy, San Jose, CA, USA, pp. 341–358. IEEE (2015)

40. Li, Y., Gai, K., Ming, Z., Zhao, H., Qiu, M.: Intercrossed access control for secure financial services on multimedia big data in cloud systems. ACM Trans. Multimedia Comput. Commun. Appl. **12**(4s), 67 (2016)

41. Mulliner, C., Robertson, W., Kirda, E.: Hidden GEMs: automated discovery of access control vulnerabilities in graphical user interfaces. In: IEEE Symposium on Security and Privacy, San Jose, CA, USA, pp. 149–162. IEEE (2014)

42. Sen, S., Guha, S., Datta, A., Rajamani, S., Tsai, J., Wing, J.: Bootstrapping privacy compliance in big data systems. In: IEEE Symposium on Security and Privacy, San Jose, CA, USA, pp. 327–342. IEEE (2014)

43. Vilk, J., Molnar, D., Livshits, B., Ofek, E., Rossbach, C., Moshchuk, A., Wang, H.J., Gal, R.: SurroundWeb: mitigating privacy concerns in a 3D web browser. In: IEEE Symposium on Security and Privacy, San Jose, CA, USA, pp. 431–446. IEEE (2015)

44. Zhu, L., Hu, Z., Heidemann, J., Wessels, D., Mankin, A., Somaiya, N.: Connection-oriented DNS to improve privacy and security. In: IEEE Symposium on Security and Privacy, San Jose, CA, USA, pp. 171–186. IEEE (2015)

45. Mehmood, A., et al.: Protection of big data privacy. IEEE Access **4**, 1821–1834 (2016)

46. Chen, C., et al.: An efficient privacy-preserving ranked keyword search method. IEEE Trans. Parallel Distrib. Syst. **27**(4), 951–963 (2016)

47. Samuel, A., et al.: A framework for composition and enforcement of privacy-aware and context-driven authorization mechanism for multimedia big data. IEEE Trans. Multimedia **17**(9), 1484–1494 (2015)
48. Gai, K., Qiu, M., Zhao, H.: Privacy-Preserving Data Encryption Strategy for Big Data in Mobile Cloud Computing. IEEE Transactions on Big Data. IEEE (2016)
49. Li, H., Wang, K., Liu, X., Sun, Y., Guo, S.: A Selective Privacy-Preserving Approach for Multimedia Data. Cyber Security. IEEE (2017)

Smart Solar Energy Based Irrigation System with GSM

C. Bhuvaneswari[1], K. Vasanth[2], S. M. Shyni[1], and S. Saravanan[3(✉)]

[1] Department of E.E.E., Sathyabama University, Chennai, India
[2] Department of E.C.E., Vidya Jyothi Institute of Technology, Hyderabad, India
vasanthecek@gmail.com
[3] Department of E.C.E., Jeppiaar Maamallan Engineering College,
Kanchipuram, India
saravanantec@yahoo.co.in

Abstract. A solar energy based Smart irrigation system has been proposed. The agricultural land is monitored with the help of three sensors and irrigation is automated with the help of Microcontroller. The status of the irrigation of the particular land is intimated to the users Mobile using GSM Technique. By using this irrigation system, resources like water and Electricity can be conservatively used with increase in irrigation efficiency, and reduction in labour cost.

Keywords: Smart irrigation system · Water level sensor · Soil moisture
GSM · Low cost system

1 Introduction

Agriculture is the backbone of a Country's Economy. In India, agriculture is mostly done manually. Irrigation is the Process of watering Plants artificially. Irrigation depends upon Variety of climate, Irregular Monsoon, Character of Soil, High Breed seeds and many other factors. In the recent past, researchers have given various automatic methods of irrigation which have been established in many countries. Few papers have been studied and described below.

An automated irrigation system with Bluetooth wireless technology for sensing and control of agricultural systems has been initiated. An embedded system with a pro-grammable logic controller is used to store the details of crop and the soil with low cost wireless radio communication [1]. Norocel Codreanu gives a comparative measure-ment with the data acquisition technique using two thermocouples. The measurement was done on the first and thirtieth day and the performance was measured [2]. Venkata Naga Rohit Gunturi discusses about the automatic irrigation which gives an interrupt signal to the sprinkler for activation whenever there are changes in the temperature. This configuration helps in saving money and water [3]. This paper proposes a fuzzy logic controller which switches the pump on or off based on the sensor readings. Various Parameters like soil moisture, leaf wetness, temperature, relative humidity, plant root depth, soil texture and water storage capacity of soil are taken into con-sideration [4]. Automatic Irrigation is done to specified areas in Pakistan to control the wastage of water. Wireless sensor Network is used for proper monitoring of the fields,

© Springer Nature Singapore Pte Ltd. 2019
L. Akoglu et al. (Eds.): ICIIT 2018, CCIS 941, pp. 75–85, 2019.
https://doi.org/10.1007/978-981-13-3582-2_6

low cost and labour, accurate and fast decision making [5, 7]. The Smart Irrigation system is implemented with Fuzzy Logic Controller. This method is applied for Precision Agriculture like Croplands or Nursery [6]. The smart irrigation system is done using solar energy and GSM Facility. Also a laser based fencing system is used for security purposes [8]. The paper proposes four soil moisture sensors for four different crops which are connected to the base station. The communication is done by X-bee PRO which reduces the cost also to a remarkable extent [9]. The author of [8] proposes an automatic monitoring system using wireless monitoring network to improve the efficiency of the irrigation system. Various sensors like soil, Humidity, Temperature, Pressure Regulator, Molecular Sensors have been used [10]. A ZigBee based water pumping system has been implemented by the author Mohit Bansal. Different Parameters have been compared with IEEE 802.11b and Bluetooth technologies with ZigBee [11]. The author Jyothipriya has designed an embedded system that would control the valve of an automated solenoid to turn on and off of a main gate valve for a specified time. This is done with a ZigBee Microcontroller informing the status to the user's mobile [12]. A Wireless data acquisition network has been used to irrigate trees. The cost and the efficiency for different types of irrigation like drip irrigation, sprinkler irrigation and surface irrigation have been compared [13]. The author Prof. Kathale and Bhaltadak1 gives a detailed survey of various irrigation methods and its control [14, 15]. The implementation of the proper system also monitors and controls the green house environment [15].

This paper proposes a three sensor based system with GSM Facility. The Sect. 1 Consists of Introduction, Sect. 2 Consists of Architecture, Sects. 3, 4, 5 and 6 consists of GSM Based Smart Irrigation System, Implementation of the irrigation system, technical specifications and Conclusion respectively.

2 Proposed Architecture for Smart Irrigation

The Architecture of the system is illustrated in Fig. 1. It Consists of the Photo Voltaic Tracking Device, Sensors and the Microcontroller.

2.1 Photo Voltaic Tracking Device

The Photo voltaic Tracking Device Consists of a Photo Voltaic Cell which converts solar energy into Electrical Energy. The Silicon Panel used produces 5 W of Power with a Maximum current of 400 mA. A flat panel is used to minimise the angle of incidence between the Sunlight and the panel. The Power Produced by the Panel is given to the Voltage Regulator by the Charge Controller and the Relay Circuit. The relay is used to switch on and off the PV Panel to the main circuit depending on the intensity of the sunlight. During the day, when the intensity of the sunlight is high, the relay is switched on to the battery for charging and to the voltage regulator to run the system. During the night, when the intensity of the sunlight is low, the relay is switched off and the circuit receives power from the battery. The tracking device with the help of Light dependent resistor rotates the panel to the sun rays to get maximum power. The Light Dependent Resistors helps in sensing the intensity of sun rays.

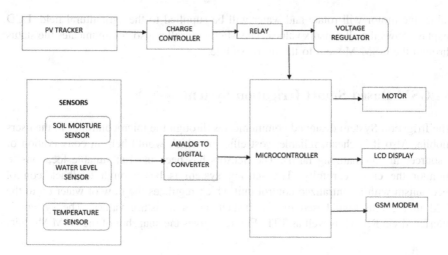

Fig. 1. Architecture of the irrigation system

2.2 Sensors

The Three Sensors are used namely Soil Moisture Sensor, Water level Sensor and Temperature Sensor. Moisture sensor sense the amount of moisture in the soil, temperature sensor measure the temperature and the water level sensor check the level of water in the field. A soil moisture probe is made up of multiple soil moisture sensors. Since analytical measurement of free soil moisture requires removing a sample and drying it to extract moisture, soil moisture sensors measure some other property, such as electrical resistance, dielectric constant, or interaction with neutrons, as a proxy for moisture content. The water level sensor is connected with the Microcontroller and it is made up of floating type of plastic, which floats in water to sense the level of water. When the water is full in the land the floating type sensor will float in water and reaches the top edge which used to indicate the water is full. Also, when water is low in the land it reaches the bottom and indicates the microcontroller that water is low. The Temperature Sensor uses a Thermistor which changes its resistance with change in temperature. The operating temperature range is from −55 °C to 150 °C and the output voltage varies by 10 mV in response to every degree Celsius rise/fall in ambient temperature.

2.3 Microcontroller

An 8051 Microcontroller is used in conjunction with an Analog to Digital Converter. The Soil Moisture Sensor, Water Level Sensor and the Temperature Sensor gives the analog output to the Analog to Digital Converter. The Digital Output of the Analog to Digital Converter is fed to the Microcontroller for giving instructions to the output circuit. The Microcontroller is connected with the Motor, LCD Display and GSM Modem. The rotation of the Motor depends on the value of the sensors communicated through the Microcontroller. If the sensors shows normal value, then the motor will be in stop condition and it will not rotate. If anyone sensor exceeds the given normal value, the motor will rotate and water will be supplied to the agricultural field. LCD display shows the all sensors values. The Microcontroller also communicates the status through the GSM Modem to the users Mobile.

3 GSM Based Smart Irrigation System

The Irrigation System designed communicates through the microcontroller to the users mobile. Also it is cheap, reliable, cost efficient which would help in conservation of resources in automatizing farms. The Sensors has been used at suitable locations to monitor the crops carefully. The sensing system is based on a feedback control mechanism with a centralized control unit which regulates the flow of water on to the field in the real time based on the instantaneous moisture values. The system is interfaced via RS232 as well as TTL. Figure 2 gives the snapshot of the GSM Module.

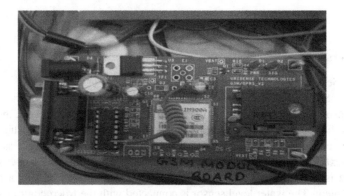

Fig. 2. GSM module

Initially all the sensor values are read and assigned to port3. Three sensors have been used namely sensor 1 (Temperature Sensor), Sensor 2 (Water Level Sensor), and Sensor 3 (Soil Moisture Sensor). If Sensor 1 has a value greater than 40, then the temperature is displayed as high on the LCD. If Sensor 2 value is less than 50, then F0 is displayed in the LCD. Also if sensor 3 reads a value less than 150, then the moisture level is displayed as low. Accordingly the motor is brought to a stop condition. Time Delay is set to 250 ms. The operation is represented in different Modes stated in Fig. 3 and Table 1.

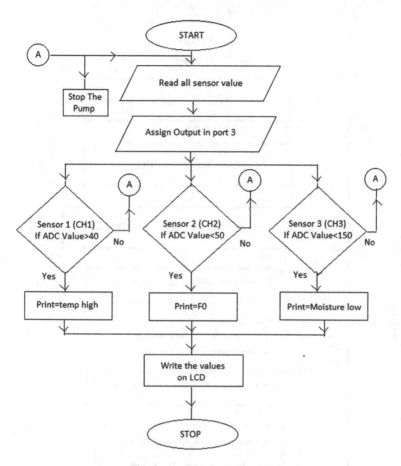

Fig. 3. Process flow diagram

Table 1. Modes of operation

Modes	Sensor 1	Sensor 2	Sensor 3
Temperature mode	On	Off	Off
Water level mode	Off	On	Off
Moisture mode	Off	Off	On
All the above	On	On	On

The operation of LCD has been explained in Fig. 4 Flowchart. The three sensors reads the sensor values and gives to the Analog to Digital Converter (ADC 0808) for conversion. When received input from the sensors, the ADC Communicates to the microcontroller through the input and output ports which in turn gives to the display unit and the motor. The process is represented diagrammatically in Fig. 4.

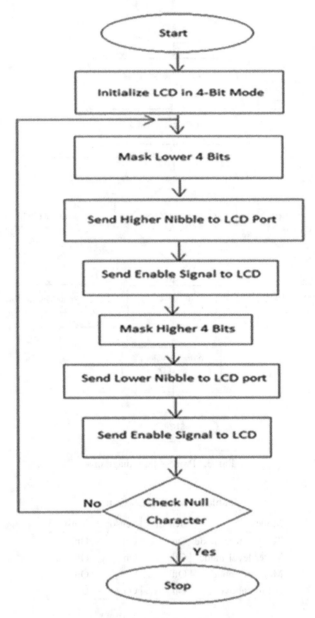

Fig. 4. LCD flowchart

4   Implementation of Irrigation System

The Smart irrigation system is simulated using Proteus 7.6 and KEIL COMPILER for 8051. Microcontroller 8051 is interfaced with ADC0808 Analog to digital Converter whose ports are connected to the temperature sensor, water level sensor and Soil Moisture Sensor. The program coding is given to the Microcontroller which gives instructions to the three sensors. Water level Sensor is connected with Port 2, Soil Moisture Sensor is connected with Port 1 and temperature sensor is connected to Port 28 with ADC0808. In turn the Microcontroller is connected to a motor, which gives instructions for motor to rotate or to stop and LCD Display to inform the status of the field. Figure 6 gives the simulation circuit and Fig. 7 gives the Simulation Circuit with sensor values (Fig. 5).

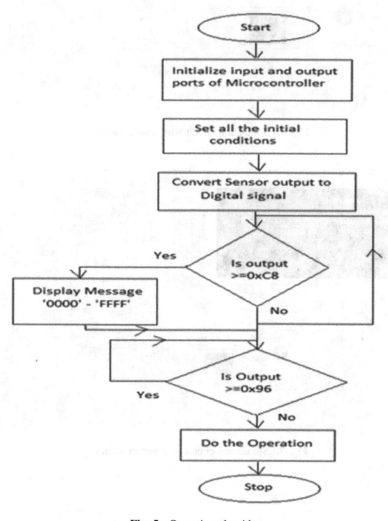

Fig. 5. Operating algorithm

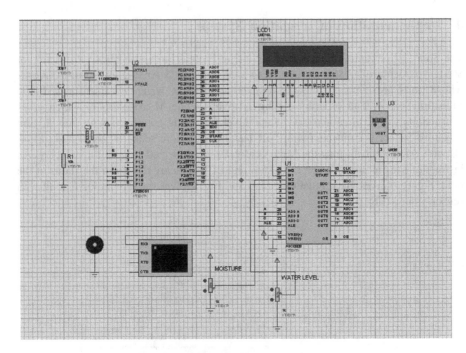

Fig. 6. Simulation circuit with sensor values

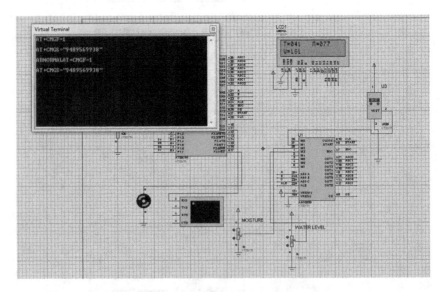

Fig. 7. Simulation circuit with sensor values

5   Technical Specifications

The Solar Panel used occupies an area of 175 w/sq meter with a weight of 15.8 kg/sq meter. The Battery used is 12 V, 2.5 AH Lead-Acid battery having a weight of 0.91 kg. A 32 Pin 8051 Microcontroller is used in conjunction with ULN2003 Motor Driver and ADC0808 Analog to Digital Converter. The different types of sensors used are Soil Moisture Sensor with voltage range of 3.3 to 5 V with dimension of $5.0 \times 2.6 \times 1.7$ cm. IC LM35 is the temperature sensor used. All other component specifications has been listed out. Figure 8 shows the hardware of the smart irrigation system and the specifications are stated in Table 2.

Table 2. Technical specifications

Components	Specifications
Transformer	230 V AC/12-0-12-0 AC
Rectifier circuit	12 V AC-12 V DC
Pumping motor	12 V, 0.5 A
Analog to digital converter	ADC0808
Tracking device motor	23 V, 0.65 A

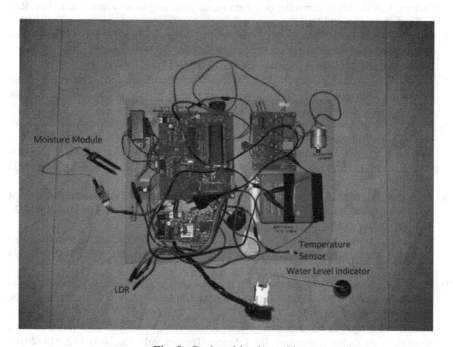

Fig. 8. Designed hardware kit

6 Conclusion

There has been many causes of low productivity in India like population on land, use of Manures, lack of adequate finance, outdated agricultural techniques, absence of Productive investment, etc., These causes can to some extent be overcome by using smart irrigation techniques proposed. The Proposed method will help to reduce the cost, reduce the manual intervention and also helps in higher productivity. As many Parts of our country has water scarcity Problem, this method can give a good solution. Also, this method is less costly and can solve the water problems for irrigation.

References

1. Sandhu, S.S., Gupta, R., Garg, R.: Smart irrigation system using wireless communication. In: Proceedings of the International Conference on Advances in Electronics, Electrical and Computer Science Engineering — EEC 2012. Vit University, Vellore (2012). https://doi.org/10.3850/978-981-07-2950-9. ISBN 978-981-07-2950-9
2. Codreanu, N., Varzaru, G., Ionescu, C.: Solar powered wireless multi–sensor device for an irrigation system. In: 37th International Spring Seminar on Electronics Technology, Center for Electronic Technology and Interconnection Techniques, Politehnica University, Bucharest, Romania. IEEE (2014)
3. Gunturi, V.N.R.: Micro controller based automatic plant irrigation system. Int. J. Adv. Res. Technol. 2(4), 194–198 (2013)
4. Patil, P., Kulkarni, U., Desai, B.L., Benagi, V.I., Naragund, V.B.: Fuzzy logic based irrigation control system using wireless sensor network for precision agriculture. In: Agro-Informatics and Precision Agriculture (AIPA 2012) (2012)
5. Jadoon, S., Solehria, S.F., Qayum, M.: A proposed least cost framework of irrigation control system based on sensor network for efficient water management in Pakistan. Int. J. Basic Appl. Sci. (IJBAS-IJENS) 11(2), 11 (2011)
6. Patil, P., Desai, B.L.: Intelligent irrigation control system by employing wireless sensor networks. Int. J. Comput. Appl. 79(11), 33–40 (2013). (0975 – 8887)
7. Kansara, K., Zaveri, V., Shah, S., Delwadkar, S., Jani, K.: Sensor based automated irrigation system with IOT: a technical review. Int. J. Comput. Sci. Inf. Technol. 6(6), 5331–5333 (2015)
8. Babu Rajendra Prasad, D., Sunil, N., Drupad, V., Madhu, C.N., Yashavantha, B.K.: GSM based smart agriculture system with auto solar tracking. Int. J. Electr. Electron. Res. 3(2), 392–397 (2015). ISSN 2348-6988 (online)
9. Kaushik, M., Devi, A., Ratan, R., Luthra, S.K.: Remotely controlled microcontroller based automatic irrigation system. Int. J. Latest Trends Eng. Technol. (IJLTET) 5(4), 269–274 (2015)
10. Hade, A.H., Sengupta, M.K.: Automatic control of drip irrigation system & monitoring of soil by wireless. IOSR J. Agric. Veterinary Sci. (IOSR-JAVS) 7(4), 57–61 (2014). e-ISSN: 2319-2380, p-ISSN: 2319-2372, Ver. III
11. Bansal, M., Bhatia, T., Srivastava, S., Gupta, S., Goyal, T.: Automatic solar powered water pumping using ZigBee technology. Int. J. Emerg. Technol. Adv. Eng. 4(4), 812–816 (2014). ISSN 2250-2459, ISO 9001:2008 Certified Journal

12. Jyothipriya, A.N., Saravanabava, T.P.: Design of embedded systems for drip irrigation automation. Int. J. Eng. Sci. Invention 2(4), 34–37 (2013). ISSN (Online): 2319 – 6734, ISSN (Print): 2319 – 6726

13. Aarthi, R., Shaik Abdul Khadir, A.: A wireless application of automatic control of drip irrigation system. Int. J. Adv. Res. Comput. Commun. Eng. 4(10), 499–502 (2015)

14. Kathale, P.P., Mankari, J., Shire, P.: A review on monitoring and controlling system for the operation of greenhouse environment. Int. J. Adv. Res. Comput. Sci. Softw. Eng. 5(4) (2015). ISSN 2277 128X

15. Bhaltadak, N.S., Ingale, H.T., Chaudhari, S.K.: GSM based remote sensing and control of an irrigation system using WSN: a survey. Int. J. Innov. Res. Sci. Eng. Technol. (An ISO 3297: 2007 Certified Organization) 4(6) (2015)

Data Analytics and its Applications

Understanding the Role of Visual Features in Emoji Similarity

Sunny Rai[1](✉), Apar Garg[2], and Shampa Chakraverty[2]

[1] School of Engineering Sciences, Mahindra École Centrale, Hyderabad, India
`sunny.rai@mechyd.ac.in`
[2] Division of Computer Engineering,
Netaji Subhas Institute Technology, Delhi, India
`apargarg05@gmail.com, apmahs.nsit@gmail.com`

Abstract. Emojis are pictographs which provide emotional cues and creativity to an otherwise bland textual conversation. They are widely used across different social media platforms to express ineffable feelings and facilitate an intimate conversation. The extent of its popularity can be gauged from the growing number of emojis in upcoming Emoji version. However, prior works on emoji prediction have majorly emphasized semantic relatedness. In this paper, we attempt to understand the significance of visual similarity and thus, the contribution of visual features in computing similarity of emojis. We use a publicly available dataset *EmoSim508* to perform our experiments. The results indicate a correlation between visual features and emoji similarity.

1   Introduction

Humans are expressive beings who employ different literary tools such as metaphors along with visual and auditory cues to subtly convey their feelings and opinions. After the advent of social media platforms, emoticons and emojis crept into our digital vocabulary to compensate for the lack of visual gestures and expressions. The Japanese word *Emoji* is composed of words, *e* (picture) + *mo* (writing) + *ji* (character). It is a set of pictographs which represent facial expressions as well as entities and events from our regular lives. The Emoji Version 11.0 further expands the existing set by adding 157 new emojis[1]. The growing size and popularity of emojis corroborate their significance in making our digital conversations more lively and personal [1,2].

Due to fluid and ambiguous nature of these pictographs [3,4], it is often difficult to organize them on a digital interface. In [5], authors clearly illustrate the problems faced by a user while selecting an emoji on Google scrollable screen keyboard. Recently, *Swiftkey* keyboard introduced a new feature, *emoji prediction* which suggests 18 emojis in accordance with the typed text[2]. *Minuum* is

---

[1] https://blog.emojipedia.org/157-new-emojis-in-the-final-2018-emoji-list/.
[2] https://goo.gl/BdN1L1 accessed on March 20, 2018.

© Springer Nature Singapore Pte Ltd. 2019
L. Akoglu et al. (Eds.): ICIIT 2018, CCIS 941, pp. 89–97, 2019.
https://doi.org/10.1007/978-981-13-3582-2_7

another keyboard which provides support for smart emoji prediction[3]. These predictions are fundamentally based on the semantic similarity of the typed text with the available set of emojis.

Prior works on emoji similarity have utilized either categorical similarity such as Jaccard coefficient which is computed on a set of words extracted from emoji description, or semantic similarity calculated using vector representations of candidate emojis [5–8]. However, we observe that categorical similarity measures often fail to capture the contextual similarity that is, relation between emojis which occurs in similar contexts and thus, it is incapable of identifying latent relatedness between them. In contrast, word embeddings are known to encapsulate diverse forms of similarity between two emojis. This includes situational (🍴🍽) as well as emotional similarity (💔😢, 😔😞). However, we note that the word embeddings are trained using the notion of distributional similarity of words and therefore, tend to miss visual relatedness between emojis. For example, while predicting similar emojis for 👣 on the basis of categorical similarity, authors in [5] obtain {👣 ⟶ 👋, 👅, 👂} whereas for semantic similarity, the obtained set of similar emojis is {👣 ⟶ 📊, ♻, 📳}. Interestingly, we note that categorical similarity interprets 👣 as a body part and thus, suggests *hand, tongue* and *ear* as its similar emojis. Whereas word embeddings understand 👣 in context of pollution and thus as carbon footprints. Consequently, it suggests *barplot* (possibly indicating increasing pollution or temperature), *recycling* and *vibration* (possibly to indicate radiation).

In this paper, we attempt to understand the impact of visual features in determining emoji similarity. Till now, the work on emoji similarity has emphasized on capturing similarity from the textual descriptions of emojis. But these descriptions do not usually provide the visual description for emojis. Thus, the captured similarity only reflects the semantic relatedness for a given pair of emojis.

The EmoSim508 dataset has human annotated similarity scores for a pair of emojis on the basis of two criteria [8] that are,

(a) How equivalent are the given two emojis? (i.e., can the use of one emoji be replaced by the other?) and
(b) How related are the given two emojis? (i.e., can one use either emoji in the same context?)

In the former question, the pair {🏠 💔} has a score of 0.4 whereas in the latter question, its score is 1.6. Here, the pair {🏠 💔} has significantly lower similarity than relatedness because sadness or crying is not always because of distress from romantic reasons such as unrequited love. So, 💔 is not a suitable substitute for 🏠. Likewise, for {🎉 🎂}, the score is 0.9 for (a) whereas 2.6 for (b), indicating that 🎂 does not always connote a party (such as birthday celebration). Thus, it can be said that a user assess the visual similarity of emojis for question (a) whereas (b) is more about contextual relatedness.

---

[3] https://goo.gl/z8bex5 accessed on March 20, 2018.

In this paper, we utilize a bag of visual words approach to quantify the visual similarity between two emojis. We use Scale Invariant Feature Transform (SIFT) [9] from OpenCV library [10] to extract image descriptors. It serves as a bag of visual words on which basis we generate vector representations for emoji images. We use Spearman's rank correlation to evaluate the efficacy of generated visual embeddings in detecting emoji similarity for instances in EmoSim508 dataset.

The rest of the paper is organized as follows. In Sect. 2, we provide a brief introduction to prior works on the semantic similarity of emojis. We explain the feature extraction and thus, its effect on computing emoji similarity in Sect. 3. This is followed by the conclusion and a direction for future work.

2   Related Work

The role of emojis in digital communication has been a subject of interest for a long time. With its proven significance in building connections on digital platforms [11], we see a surge in interest of computational linguists in understanding the relevance of emojis in human communications. Authors in [12] also show the user's preference for emojis over relatively textual emoticons.

Majority of the researchers have focused on generating emoji embedding either by using pre-trained *word2vec* embeddings or by training their model on a corpus having emojis such as Twitter. Barbieri *et al.* [6] and Pohl *et al.* [5] used the skip-gram model to train emoji embeddings on English tweets. The generated embeddings are capable of segregating similar emojis however, it has limited coverage as users tend to use only a small set of emojis in their daily conversations. Eisner *et al.* [7] used emoji description as corpus for training their model. They compute the sum of individual pre-trained *word2vec* vectors in the emoji description to generate the emoji embedding. Ai *et al.* [13] used the input of 1.22 billion messages from Kika Emoji Keyboard. They highlight the correlation between the sentimental context and emoji usage. However, they also note the power-law distribution (zipf's distribution) from uneven emoji's usage. Wijeratne *et al.* [8] extracted emoji descriptions, emoji sense labels and emoji sense description from EmojiNet [14] to train emoji embeddings. Similar to the approach proposed in [7], they also used pre-trained *word2vec* embeddings to represent words in the extracted description from EmojiNet. Their approach performed the best when trained on emoji sense label. They also released *EmoSim508* dataset which contains 508 emoji pairs with human annotated similarity scores.

3   Understanding the Role of Visual Features for Emoji Similarity

3.1   Problem Statement

In this paper, we analyze the effect of visual features in determining the similarity between two emojis. We use EmoSim508 dataset [8] in which every instance comprises of a pair of emojis and the human annotated ratings for three criteria dealing with *similarity*, *relatedness* and *mean of the first two ratings*. These ratings are discussed in detail in Sect. 3.4.

3.2 Building Training Dataset

For training, we build the corpus by extracting images from Full Emoji list v11.0 available on Unicode website[4]. We consider images from 5 vendors to incorporate representational diversity across different platforms. The considered vendors are *Apple, Google, Twitter, EmojiOne* and *Facebook* which have significant presence across digital devices. We only consider emojis which are available on all the mentioned platforms. Therefore, the total number of unique emojis in the dataset is 1559 and the total number of images is 7795. We illustrate *hugging face* emoji for all five vendors in Fig. 1.

Fig. 1. Hugging Face (a) Apple (b) Google (c) Twitter (d) EmojiOne (e) Facebook

3.3 Generating Emoji Visual Embeddings

The underlying idea behind our approach is to capture visual relatedness between a given pair of emojis. To achieve that, we extract image descriptors through SIFT module in OpenCV library. The SIFT algorithm is a popular method in image processing which is invariant to intensity or rotational variations images. It identifies a set of characteristic image descriptors which serve as a bag of visual words (BOV) and help in distinguishing a particular subsection of an image from another image [15,16]. We use k-means algorithm to group similar descriptors from all the images which is analogous to clustering of synonyms in bag of word representation in textual domain. This acts as an image embedding. However, the problem is to determine the length of descriptors which succinctly retain the image information without being redundant.

To determine the size of BOV, we test the effectiveness of the learned vector on an emoji classification problem. We train a linear support vector machine model for the emoji classification using the learned visual representation with varying vector sizes. We use the metric *accuracy* to evaluate the performance of the trained classifier. It is defined as in Eq. 1.

$$Accuracy = \frac{\#Correctly\ Classified\ Images\ (emojis)}{\#Images\ in\ dataset} \tag{1}$$

We test the classifier for sizes $= \{10, 20, 40, 50, 100, 150, 200, 250, 300\}$. From Fig. 2, we can observe that classification accuracy of the model is steeply increasing as we increase the size of image embedding. However, as the size becomes 100, we observe very subtle improvement in accuracy indicating a saturation. Thus, we can conclude that 100 is the most appropriate size to consider for our problem statement.

---

[4] https://unicode.org/emoji/charts/full-emoji-list.html.

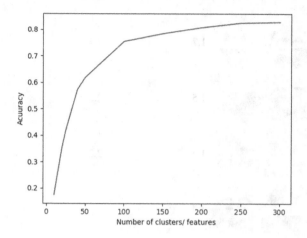

Fig. 2. Visual embedding size vs classification accuracy

3.4   Experiments and Results

The objective is to analyze the contribution of visual relatedness while computing emoji similarity. Till now, researchers have used textual embeddings to compute emoji similarity. We consider *emoji2vec* embeddings [7] and embeddings generated using EmojiNet sense labels [14] as our baselines for comparison. These are textual embeddings which exploit the notion of distributional hypothesis and thus, logically more apt to capture contextual relatedness. We use Python V2.7 and OpenCV library [10] to implement our approach.

Test Dataset. We use a publicly available EmoSim508 dataset to perform our experiments [8]. The dataset has human ratings within range 0–4 where 0 is the lowest value which can be assigned for a given case which are follows:

1. **Q1:** How equivalent are the given two emojis? (i.e., can the use of one emoji be replaced by the other?)
2. **Q2:** How related are the given two emojis? (i.e., can one use either emoji in the same context?)
3. **Mean:** How related are the given two emojis when the average of the previous two ratings is considered?

   In Fig. 3, we provide a subset a few instances with their ratings. The first column indicates instance ID followed by the emoji pair as the next two elements. The fourth column provides the human rating for **Q1** whereas the fifth column is the rating for **Q2**. The last column provides the mean value of **Q1** and **Q2** ratings.

3.5   Experiments

The *emoji2vec* embeddings are publicly available [7] and we use the same for our experiments.

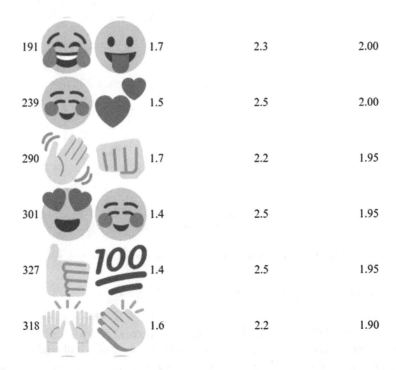

191			1.7	2.3	2.00
239			1.5	2.5	2.00
290			1.7	2.2	1.95
301			1.4	2.5	1.95
327			1.4	2.5	1.95
318			1.6	2.2	1.90

Fig. 3. A subset of EmoSim508 dataset

For **EmojiNet sense labels based emoji representation**, we implement an approach similar to the technique proposed by Wijeratne *et al.* The EmojiNet sense labels are word-PoS tag pairs which describe the senses under which an emoji can be used. For example, 🙈 (see no evil monkey) has six noun senses (*hands, monkey, pet monkey, evil, gesture* and *not*), seven verb senses (*see, blind, hiding, gesture, not, forbidden* and *prohibited*) and seven adjective senses (*forbidden, no, prohibited, scared, blind, embarrassed* and *evil*). The summation of pre-trained *word2vec* embeddings for all sense labels are taken to generate the textual emoji embedding. Unlike the approach proposed in [8], we take the sum instead of average of EmojiNet sense labels *word2vec* embeddings. The average is a form of normalization of summed up embeddings. Since cosine distance is invariant to scaling and therefore, taking average or summation of embeddings have no impact on the obtained cosine similarity.

Performance Evaluation. We used cosine similarity as the metric to compute similarity between two emoji embeddings. After that, we computed Spearman's Rank correlation between the predicted similarities and human annotated similarities to gauge the overall efficacy of the generated embeddings. We also calculated Spearman correlation on Jensen-Shannon (JS) divergence metric. Higher the JS distance, lower the similarity between the given pair of emojis. However, the results for JS divergence are equivalent to inverse of cosine similarity and therefore, we do not report it in this paper.

First, we calculated cosine similarities for our baseline embeddings that are, *emoji2vec* and *sense-label(word2vec)*. After that, we conducted experiments on generated visual embeddings in steps to analyze their contribution in determining emoji similarity. We first computed cosine similarity between visual embeddings for a given pair of emojis to detect visual similarity between them. From the results in Table 1, we observe that visual embeddings by itself do not provide a significant basis to determine similarity between two emojis. So, we trained a linear regression (LR) model where we consider cosine similarity computed using visual embeddings (F3) as well as *sense_labels* (F2) as features. This helped us in understanding if there is any impact of visual features on computing emoji similarity.

Table 1. Spearman's rank correlation for similarity task

F#	Approach	Q1	Q2	Mean
F1	*emoji2vec*	0.633	0.653	0.658
F2	sense_label(*word2vec*)	0.721	0.718	0.733
F3	Visual Embeddings	0.276	0.238	0.259
F2, F3	LR(F2,F3)	0.718	0.718	0.732

3.6 Discussion

We provide the results in Table 1. From the results, we observe that *sense_labels* trained on *word2vec* has the highest correlation in all the three cases. It has higher correlation with **Q1** in comparison with **Q2** which is interesting since *word2vec* representations are known to be more effective in capturing contextual relatedness instead of similarity.

The visual vector representation has a higher correlation for **Q1** in comparison with **Q2** which is as per our hypothesis. The LR model provides comparable performance to *sense_labels* representation.

However, on close investigation of trained LR function for **Q1**, we observe a pattern in the contribution of visual and *sense_labels* representation as shown in Table 2. The coefficient for F3 is at its highest for **Q1**, indicating that it matters the most when we consider similarity. Whereas the coefficient for F2 is at its highest for **Q2** when contextual similarity is at play. That is, visual similarity do have a role while computing the emoji similarity *i.e.* **Q1**. However, we need to study in detail about its significance.

We also note that the many instances in EmoSim508 dataset are somewhat correlated for **Q1** and **Q2**. As the similarity reduces, so does its contextual relatedness. Thus, *sense_labels* embeddings perform reasonably even for **Q1**. However, this property of the dataset makes it unsuitable in understanding the impact of visual features.

Table 2. Coefficients for features in LR model

Q#	Coefficient for F2	Coefficient for F3
Q1	0.68	0.149
Q2	0.832	0.054
MEAN	0.77	0.093

Table 3. Instances from EmoSim508

Instance-ID	Human-Ratings	sense_label	LR(F2,F3)
ID-138 ( )	0.6	1.52	0.59
ID-186 ( )	0.5	1.44	0.53

Nevertheless, as shown in Table 3, we found cases such as instance ID-138 ( ) and ID-186 ( ) where the values predicted by LR model *i.e.* 0.59 and 0.53 were more closer to their user ratings which are 0.6 and 0.5 for **Q1** than the values *i.e.* 1.52 and 1.44 obtained using scaled cosine similarity between *sense_label* representation. It should also be noted that humans often do not know the difference between yellow or green heart emojis and usually consider them similar.

4   Conclusion

In this paper, we studied the role of visual features in emoji prediction. We observed a correlation between visual features and similarity score given in EmoSim508 dataset. However, our technique is simple and needs refinement to capture the visual similarity effectively. In future, we will train a Convolutional Neural Network to generate image embeddings for emojis.

References

1. Pohl, H., Murray-Smith, R.: Focused and casual interactions: allowing users to vary their level of engagement. In: Proceedings of the SIGCHI Conference on Human Factors in Computing Systems, pp. 2223–2232. ACM (2013)
2. Janssen, J.H., Ijsselsteijn, W.A., Westerink, J.H.D.M.: How affective technologies can influence intimate interactions and improve social connectedness. Int. J. Hum. Comput. Stud. **72**(1), 33–43 (2014)
3. Cho, H., Ishida, T.: Exploring cultural differences in pictogram interpretations. In: Ishida, T. (ed.) The Language Grid. Cognitive Technologies, pp. 133–148. Springer, Heidelberg (2011). https://doi.org/10.1007/978-3-642-21178-2_9
4. Miller, H., Kluver, D., Thebault-Spieker, J., Terveen, L., Hecht, B.: Understanding emoji ambiguity in context: the role of text in emoji-related miscommunication. In: 11th International Conference on Web and Social Media, ICWSM 2017. AAAI Press (2017)

5. Pohl, H., Domin, C., Rohs, M.: Beyond just text: semantic emoji similarity modeling to support expressive communication. ACM Trans. Comput. Hum. Interact. (TOCHI) **24**(1), 6 (2017)
6. Barbieri, F., Ronzano, F., Saggion, H.: What does this emoji mean? A vector space skip-gram model for twitter emojis. In: LREC (2016)
7. Eisner, B., Rocktäschel, T., Augenstein, I., Bošnjak, M., Riedel, S.: Emoji2vec: learning emoji representations from their description (2016). arXiv preprint: arXiv:1609.08359
8. Wijeratne, S., Balasuriya, L., Sheth, A., Doran, D.: A semantics-based measure of emoji similarity (2017). arXiv preprint: arXiv:1707.04653
9. Hess, R.: An open-source siftlibrary. In: Proceedings of the 18th ACM International Conference on Multimedia, pp. 1493–1496. ACM (2010)
10. Bradski, G., Kaehler, A.: Opencv. Dr. Dobbs. J. Softw. Tools **3**, 122–125 (2000)
11. Kelly, R., Watts, L.: Characterising the inventive appropriation of emoji as relationally meaningful in mediated close personal relationships. In: Experiences of Technology Appropriation: Unanticipated Users, Usage, Circumstances, and Design (2015)
12. Pavalanathan, U., Eisenstein, J.: More emojis, less :) the competition for paralinguistic function in microblog writing. First Monday **21**(11) (2016). https://doi.org/10.5210/fm.v21i11.6879, https://firstmonday.org/ojs/index.php/fm/article/view/6879/5647. Accessed 16 Nov. 2018, ISSN 13960466
13. Ai, W., Lu, X., Liu, X., Wang, N., Huang, G., Mei, Q.: Untangling emoji popularity through semantic embeddings. In: ICWSM, pp. 2–11 (2017)
14. Wijeratne, S., Balasuriya, L., Sheth, A., Doran, D.: EmojiNet: building a machine readable sense inventory for emoji. In: Spiro, E., Ahn, Y.-Y. (eds.) SocInfo 2016, Part I. LNCS, vol. 10046, pp. 527–541. Springer, Cham (2016). https://doi.org/10.1007/978-3-319-47880-7_33
15. Lowe, D.G.: Object recognition from local scale-invariant features. In: Proceedings of the seventh IEEE International Conference on Computer Vision, vol. 2, pp. 1150–1157. IEEE (1999)
16. Lowe, D.G.: Distinctive image features from scale-invariant keypoints. Int. J. Comput. Vis. **60**(2), 91–110 (2004)

Semantic Network Based Cognitive, NLP Powered Question Answering System for Teaching Electrical Motor Concepts

Atul Prakash Prajapati[✉], Ashish Chandiok, and D. K. Chaturvedi

Faculty of Engineering, Dayalbagh Educational Institute, Agra 282005, U.P., India
atulprakash21@gmail.com, achandiok@gmail.com, dkc.foe@gmail.com
http://www.dei.ac.in

Abstract. Background. Nowadays with the advent of technology and easy access to the "WWW", there is a need for such systems that can give exact and precise answers to user's queries. It leads to the requirement of the Question-Answering System. In this Paper, we are going to present "$SN : CQA$" (Semantic Network based Cognitive Question-Answering System).

Objective. The essential purpose of designing this system is to place such systems in the remote locations (but not limited to) where internet connectivity is not yet possible (Offering off-line experiential knowledge-base). Thus students, living in remote areas, can get the benefit of existing technologies and technical education.

Methodology. For practical implementation of the proposed architecture of "$SN : CQA$", we have selected education domain (teaching "Basic Electrical Motor Concepts" to the novice users). Thus to achieve the required goal this paper proposes an architecture "$SN : CQA$", which combines the power of (i) Semantic-Network, (ii) NLP, (iii) Agent's behaviour modeling, and (iv) Cognitive behaviour modelling under one roof. "$SN : CQA$" tries to give best answers to user queries rather than the correct answers. This paper also emphasis on the requirement of experiential knowledge base (Teacher's Experiences) rather than web mining based answer's extraction.

Result. "$SN : CQA$" combines the power of several techniques under one architecture which leads to the better decision making while searching and answering the user's queries.

Conclusion. Finally this paper concludes that for the construction of an effective QA system we should go for the architectural way of designing the system, which combines the powers of many individual techniques under one roof. This paper also emphasise on the importance of experimental knowledge base.

Keywords: Question-Answering System · Semantic-Network · NLP Cognitive decision making · System approach

A. P. Prajapati and A. Chandiok are working in the field of artificial cognitive systems. D. K. Chaturvedi is working in the field of softcomputing, cognitive and conscious systems.

1   Introduction

In early days Researchers were using "Frames and Scripts" but with the enrichment of technology, people moved towards [1] "Semantic-Network" for knowledge modeling. Akin to frames and scripts this particular approach was also a feeble approach and is unable to fulfill the required objectives. Thus to provide a semantic understanding of the context one has to provide additional techniques (like NLP) to the system. Further, the concept of [2] "Ontology" had proposed. It supports and offers a range of techniques like class concept, inheritance and much more. It reuses, captures, processes, and can communicate knowledge efficiently. These days everyone is talking about Semantic-Web (do not get confused with semantic-network, both are different concepts), for gaining knowledge. Fundamentally earlier web was structured, it dealt with structured data however now a day's scenario is changed entirely. Today we have to deal with various types of data (structured, semi-structured, and unstructured). We are dealing with audio, video, text and the combination of these formats. So to handle such an unstructured data it necessitates a robust knowledge base and a robust system as well. In this section, I am going to exhibit a succinct introduction of the systems that have designed over time for the improvement of Q-A systems.

[3] proposed an automatic appraisal system to appraise the quality of answers offered by the on-line health Q-A systems. They pointed out that the existing systems are lagging behind to provide quality answers to the community questions. So, they proposed a DBN (Deep Belief Network) learning framework. [4] proposed a knowledge-graph pedestal spoken question answering system. They established that the existing speech recognition systems barely confine the acoustic features and the language models for speech recognition. Throughout this process errors and inaccuracies are ignored which mortifies the system's performance. So they suggested that the combination of NLP and speech recognizer will surmount with these problems. [5] proposed a technique for finding allied queries from the question labels and its body for the community question answering knowledge bases. To accomplish the objectives he used NER (Named Entity Recognition) and NLP (Natural Language Processing) techniques. [6] proposed a method for Text-Q-A systems, that exploit the dynamic memory networks for analysing the questions. [7] proposed a hybrid querying method (SK-Query). This proposed method prevails over the problems present in existing querying systems, which had keyword and SPARQL querying approach. [8] proposed ASLEM (Adaptive semi-supervised ELM) machine for the recognition of subjectivity of the questions asked in CQA's (Community Question-Answering). They resolute on two main problems, first one is data imbalance and the second one is labelling of data as it reasons a significant challenge while data is manually labelled. As a solution, an elucidation to this Gaussian modelling had proposed. [9] suggested an interactive dialogue based model that is for identifying customer behaviour and indulgent to the needs of the on-line customers. This system had applied to an on-line Q-A system which considers yes-no types of questions and builds a model. This model additionally forecasts the conclusions for fulfilling customer needs. [10] proposed a cognitive system

for involuntarily solving RAT (Remote Associates Test) queries. RAT problems are used to measure human's ingenuity by performing psychological tests. [11] proposed a trigger and class-based language modelling technique for getting better sentence retrieval method throughout the process of question answering. To confine the relationship in between different terms (words) they used class model. This method uses the similar concept as used in the clustering algorithm, where related words go for the same cluster. This clustering is further useful for searching germane sentences. On the other hand, the trigger model forms a word couple (target and trigger words) throughout the training process. These pairs of words had included in the trigger model, which had further used in establishing the relation in between a question and a sentence. [12] proposed a Q-A system for answering the queries of biologists. They argued that the existing system uses search engines which consult an assortment of resources for discovering the answers. Thus it is a time-consuming approach, as well as the conclusion was also not so precise. For this, they proposed a Q-A system that provides the accurate answers to the queries by consulting a variety of different type of resources. [13] proposed a keyword based searching system that applies the power of Hidden Markov model for discovering the correct resources and the power of SPARQL querying for pronouncement of the correct answer from those resources. [14] proposed a system that employs machine learning approaches (ANN and Clustering Algorithms) for discovering the right resource persons to the on-line community Q-A systems. [15] proposed an Eclipse plug-in based replay system which keeps track of the changes made throughout a software development process. This proposed system assists a programmer by providing the insights of modifications made during a software development lifecycle. [16] proposed a language independent Q-A system. The proposed system overcomes the redundancy issue which exists when we extract data from the web.

This paper consists of following sections. Introduction section gives the brief idea of the current technologies and existing systems which had designed in the area of QA systems. Background section explains the architecture of the SN : CQA. Methodology describes the implementation process of the architecture using flowcharts and algorithms. The last component of the paper experimental design and result shows the practical implementation of the SN : CQA.

2   Background

Ross Quillian in 1966 and 1968 had first defined this term "Semantic-Network". It is a graph way, $Graph\ (G := [N, E])$, to represent knowledge in the form of nodes and edges. ($N \in Nodes$) nodes in the Semantic-Network represent attributes, concepts, events, entities or values of the domain knowledge. Subsequently ($E \in Arcs$) arcs or edges represent relationship in between the attributes. Following figure ($Fig.$ 1) represents the architecture of "SN : CQA" which has following three modules $\left\{ \begin{array}{l} (i)\ SUIM,\ (ii)\ (C_b,\ RM) \in QM, \\ (iii)\ (SRM,\ NCM,\ KBM) \in TM \end{array} \right\}$.

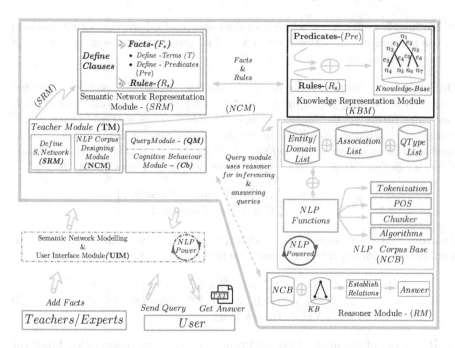

Fig. 1. Schematic diagram showing an architecture of "$SN : CQA$" a Cognitive, NLP Powered QA System.

(a) *SUIM* (*User interface module for accessing the Semantic Network*): provides an interface to access the "$SN : CQA$" system (Semantic network based NLP powered Cognitive Question Answering system). Teachers can fill the "*Concept Model*" (NCB) using (NCM) module and can create "*Facts*" and "*Rules*" (KBM) using "SRM" module. The "$SUIM$" module also provides an interface to the end users to ask questions which are handled by the "QM" module. "QM" module implicitly uses "RM" module for inferencing and answering the user queries.

(b) *TM* (*Teacher Module*): A teacher (domain expert, in "$SN : CQA$" we are consulting the teachers of the Electrical Engineering Department) uses "TM" module for building Concept model "NCB" ($NLP\,Corpus\,Base$) and Knowledge Model "KBM" ($Knowledge\,Base\,Module$). It consists of three sub modules ⟨SRM, NCM, KBM⟩.

(c) *QM* (*Query Module*): handles the user's natural language query. A user asks the query/question using user interface ($SUIM$) module subsequently this query is handled by a "QM" module which further uses "NCB" for linguistic component extraction. Linguistic components provides the required entity (E) and association (A) name. Afterwards using these tokens (E, A), "QM" module searches "KBM" for fetching the required answer from the knowledge-base. In the mean, while "QM" module uses the "RM" (Reasoner

Module) for inferencing which uses the concept of inheritance for establishing the relationship in between entities and in getting the answer. It consists of two sub-modules $\langle C_b,\ RM \rangle$.

Definition 1. *The first tuple of "TM" module is "SRM" module. A teacher uses "SRM" module for defining "Facts" and "Rules." The collection of Facts and rules forms up the knowledge base "KBM". Thus a teacher uses "TM" module and fills the knowledge in the knowledge-base (KBM) using "SRM" module. Following equations define the elements of "KBM" mathematically. First, we define "Atoms", then define "Predicates", "Facts", and "Rules".*

$$Term\,(T) := \left\{ \begin{array}{l} Atom,\ Number,\ Variable, \\ Compound\,Term\ (CT) \end{array} \right\}. \tag{1}$$

An atom consists of entities, CT (functor(atom)), Numbers (integers, float), and Variables (beginning with an upper-case letter, unknown entities).

$$\begin{aligned} Predicate\ (Pre) := \{(NA_i)\ R\ (NE_j, NE_k)\} \\ : \forall\,(i,j,k) > 0,\ \&\ (NA \in Association\,List)\,, \\ (NE\ \in\ Entity\,List) \end{aligned} \tag{2}$$

It consists of predicates that are unconditionally true. Association binds the entities or atoms which has called Predicate or Fact.

$$Fact\,(F_s) := \left\{ \sum_1^j p_j : \forall j\ =\ [1,..,m]\ \&\ m\ >\ 0 \right\}. \tag{3}$$

Facts are a collection of predicates. These are simple statements which describes the knowledge and are unconditionally true. Unlike Rules, Each fact consists of only head its body is missing.

$$Rule\,(R_s)\ :=\ \left\{ \begin{array}{l} p_1\ :\ -\ p_2, p_3, ... : \\ \qquad\qquad \{p_1,\ p_2, p_3\ \in\ [\mathrm{Pr}\,e]\} \end{array} \right\}. \tag{4}$$

Unlike facts, Rule consists of two parts (head (p_1) & body $(p_2,\ p_3, ...)$) both are separated by if clause denoted by $:\ -$. Head has a single predicate, and its body has a sequence of predicates separated by a comma (which works as a logical "&" operator). Unlike Facts, Rules are conditional, iff the body of the rule is true then overall value is true.

Definition 2. *The "NCM" module is the second tuple of the "TM" module. The "NCM" module defines "Entities/Domain Identification KeyWords", "Associations", "Qtype" list and the "NLP functions" (Tokenization, POS (Part of Speech Tagging), Chunker, and Algorithms).*

$$N_{SE} = \left\{ \sum_{i=1}^{n} NE_i,\ \forall(n > 0) \right\} \tag{5}$$

"N_{SE}" is a set of entities that are stored in NLP Corpus-Base (NCB) entity list.

$$N_{SA} = \left\{ \sum_{J=1}^{n} NA_j \,, \, \forall (n > 0) \right\} \qquad (6)$$

"N_{SA}" is a set of associations that are stored in NLP Corpus-Base (NCB) association list.

$$N_{SDE} = \left\{ \sum_{l=1}^{n} NDE_l \,, \, \forall (n > 0) \right\} \qquad (7)$$

"N_{SDE}" represents the set of domain identification elements. "$SN : CQA$" deals with only "Factoid Type" questions which are related to the "Electrical Motor" domain. Thus to filter out rest type of questions we have implemented this concept.

$$N_{SFN} = \left\{ \sum_{o=1}^{n} NFN_o \,, \, \forall (n > 0) \right\} \qquad (8)$$

"N_{SFN}" is a set of NLP functions, which consists of following features (Split Sentence, Tokenization, POS (Parts of speech tagging), Chunker etc.)

Definition 3. *The "QM" module handles user queries/questions. It holds two sub-modules (C_b, RM). Here "C_b" module implements the cognitive behavior by using liking value function. Liking value provides the feedback to the knowledge engineer about the liking of answers from the group of users.*

Definition 4. *"C_b" uses "LVM" (Liking Value Module) engine which consists of two tuples $\langle SL, f(L) \rangle$.*

$$(SL) := \left\{ \sum_{i=-10}^{10} l_i, \forall i = (-10, ., 0, ., 10) \right\} \qquad (9)$$

"SL" is a set of liking values. For every answer provided by "$SN : CQA$", each user may assign different like values to the answers according to its perception. Based on this concept we have provided a range of liking values which ranges between $(-10, ., 0, ., 10)$ (less than zero represents "*Dissatisfied*" and greater than zero represents "*Satisfied*").

$$f(L) := \left\{ \begin{array}{l} Ans_j \sum_{i=-10}^{10} l_i, \forall (j > 0) \,, \\ \forall i = (-10, ., 0, ., 10) \end{array} \right\} \qquad (10)$$

$f(L)$ is a function which assigns a particular like value in the range between $(-10, ., 0, ., 10)$ (assigned by the user) to each answer. Whenever a user asks a query to the "$SN : CQA$", it provides the best answer by exploring its knowledge-base. Subsequently, the system asks the user to grade the particular answer based on its liking about the answer. Now, this $f(L)$ function handles this value and stores it in the form of a matrix in a separate "XML" file. This log file helps the knowledge engineer while making assumptions about the group of the "$SN : CQA$" users. For example; the knowledge engineer fills the best answers in the knowledge-base based on its knowledge, understanding, maturity, and experience. But the student may find the filled answers either complicated or straightforward. Therefore this matrix helps the knowledge engineer while making assumptions about the group of "$SN : CQA$" users. So that he can fill the answers correspondingly in the future.

$$\left\{ \begin{matrix} Reasoner\,(R_r) := uses\ inheritance\ for\ inferencing\ and\ finding \\ relations\ in\ between\ the\ facts. \end{matrix} \right\} \quad (11)$$

3 Methodology

This section of the paper explains the construction process of $SN : CQA$ by using flowcharts and algorithms.

3.1 Understanding of Question and Linguistic Component Extraction

The power of a "QA" system relies on its question understanding power. If the system can identify exact semantics of the question, it will get the best results from the knowledge corpus. Thus we are implementing "NLP" concept for the better understanding of the user's natural language question. We have used the separate corpus for implementing the "NLP" concepts. "$Equation\,5$" represents the set of "$Entities$" (entities represent the real world object like "$Induction\ Motor$"). "$Equation\,6$" represents the set of "$Associations$" which relates the entities and forms up the predicates. "$Equation\,7$" represents the set of Domain identification elements (As we are dealing with questions of basic "$Electrical\ Motor\ Concepts$" only so we have to filter out rest type of questions). "$Equation\,8$" represents the set of "NLP" functions ($Tokenization, Part\,of\,Speech\,Tagging,\ and\,Chunking$).

Example 1. Following example shows the power of NLP in the understanding of the semantics of the question. The last and main function of NLP corpus base is "$Chunking$"; it helps in the collection of linguistic components. Chunker groups the tokens in the form of NP (Noun Phrase) and VP (Verb Phrase). Afterwards, By applying some functions and algorithms, we can collect the required linguistic components (E, A) which further works as keywords while searching an answer in knowledge-base.

Steps:

- *Ask Question:*
  What is the definition of an induction motor?

- *Tokenization:*
  [*What is the definition of an induction motor*]

- *Part of Speech Tagging (POS):*
  $$\begin{bmatrix} What/WP \ is/VBZ \ the/DT \ definition/NN \\ of/IN \ an/DT \ induction/NN \ motor/NN \end{bmatrix}$$

- *Chunking:*
  $$\begin{bmatrix} [NP \ What/WP], \ [VP \ is/VBZ], \\ [NP \ the/DT \ definition/NN], \ [PP \ of/IN], \\ [NP \ an/DT \ induction/NN \ motor/NN] \end{bmatrix}$$

- *Apply Algorithms and Functions and Extract the*
  Linguistic Components (E, A) from the above Tokens.

- *Select (E): "Induction Motor".*
- *Select (A): "Definition".*
- *Search "Knowledge-Base" using (E, A) pair.*

3.2   Querying and Answer Extraction from the Knowledge-Base

Following (*Fig.* 3) explains the logical representation of the knowledge-base.

Subsequently, flowchart, refer (*Fig.* 2), explains the construction process of NCB (*Concept Model*), KBM (*Experience Model*) or (*Knowledge Model*), QM (*Querying Model*), and the implementation of LVM *Model*, *Liking ValueFunction*. First, we collect the domain information. Afterwards, using the "TM" module a knowledge engineer (here we are promoting the Experiential Knowledge-Base concept, so in this case, a teacher fills the required facts in the knowledge-base) fills the facts in the knowledge-base. Subsequently, we prepare the *Concept Model* (NCB) which helps in the process of user's natural language question understanding. Both (NCB, KBM) the corpus are ready now, so we can move towards querying process directly. A user may ask its query using natural language (*English*) (We are going to teach the concepts of "Basic Electrical Motors," so the user is allowed, only to ask questions related to the "Electrical motors"). To constrain the user's question we have implemented domain identification elements list implemented in the "NCB" module. This "NL" question is handled by the $SUIM$ interface module which has the power of natural language questions understanding (provided by "NCB"). Subsequently we extract the linguistic components (E, A) pair from the NL question. Now we fill the question template [E, A, O]. This template helps in judging the correctness of

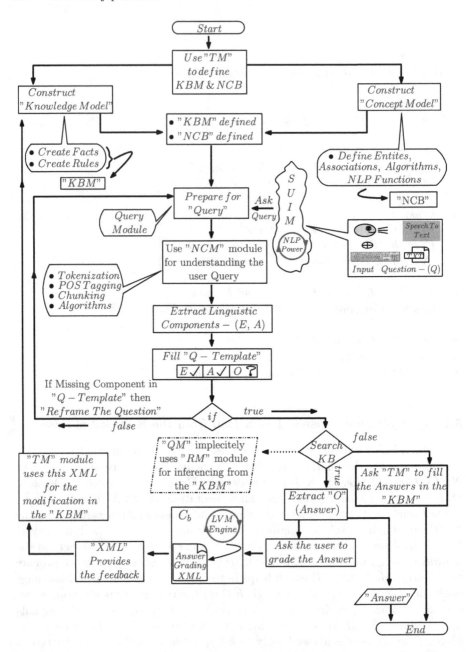

Fig. 2. Flowchart representing the "Construction and Querying Operation" of "*SN* : *CQA*"

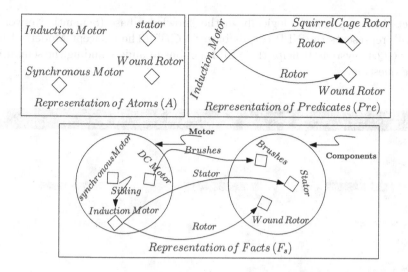

Fig. 3. Schematic diagram representing the logical description of Knowledge-base.

the question (if all the tokens of the template have filled; the question is correct otherwise it shows the error message "Re-frame the question"). Now, these tokens work as a keyword while searching in the knowledge-base. If the answer exists in the knowledge-base of "$SN : CQA$" returns the answer "O" (last token of the template which is empty initially). Otherwise, ask the "TM" (teacher) to fill the required facts in the knowledge-base. Here we have implemented the concept of feedback. For each answered question "$SN : CQA$" asks the users to rate the answer on the scale of $[-10, ., 0, ., 10]$. This value has stored in a separate "XML" file which maintains a matrix of liking values for each answer. It helps the teacher while making assumptions about the users of "$SN : CQA$".

4   Experimental Design and Result

For the implementation of "$SN : CQA$" system we have used "$C\#$" language and "Visual Studio IDE", and for the knowledge-base modeling, we have used "RDF/XML" syntax (for representing the facts). Further to implement the NLP concepts we have used SharpNLP (open source, English language natural language processing tools written in "$C\#$"). It is centred around a port of the Java OpenNLP library (http://opennlp.sourceforge.net/), which includes a sentence splitter, a tokenizer, a part-of-speech tagger, a chunker, a parser, a name finder, and consists of coreference tools.

4.1   Representation of "Concept Model" and "Knowledge Model"

Above figure "*Fig.* 3" represents the concept of knowledge modelling. It symbolizes how individual atoms (entities), predicates (*Association* (*Entity*, *Entity*)) or facts have represented logically in the knowledge-base (Semantic-Network). "*Fig.* 4" represents the NLP corpus base "*NCB*" (The Concept Model) of the "*SN : CQA*" system. It helps the NLP functions in understanding the semantics of the question.

Fig. 4. Schematic diagram representing the "Concept Model - (NCB)" of "*SN : CQA*"

4.2   Representation of Cognitive QA System "SN: CQA"

"*Figure* 5" represents the practical implementation of the "*SN : CQA*" a cognitive QA system. A user can ask the question by typing from the keyboard which is handled by "*QM*" module (as the speech processing will be implemented soon). The "*QM*" module uses the NLP function, which implicitly uses the "*NCB*", for understanding the semantics of the question and it extracts the linguistics components. These tokens work as keywords while searching in the knowledge-base (Semantic-Network). The "*QM*" module implicitly uses the "*RM*" module for inferencing. When the system finds the answer to the user question, it asks to provide the feedback (based on the liking of the answer) from the user by grading the response on $(-10, ., 0, ., 10)$ scale. This feedback is stored separately in an *"XML"* file which maintains a matrix of liking values associated with each solution. This *"XML"* file further helps the domain experts in knowledge updation and user's behavior modeling.

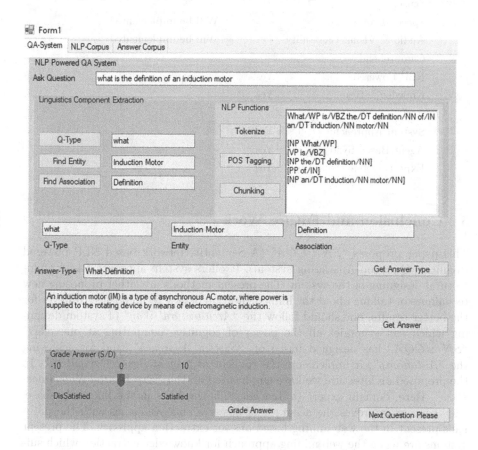

Fig. 5. Schematic diagram representing the interface of *"SN : CQA"*, a Cognitive Q-A System.

4.3   Cognitive Computing Capabilities of SN: CQA

Cognitive computing defines the way to synergize the various techniques under one roof. One should follow the architectural approach of system designing if he/she wants to implement the full power of cognitive computing. [17], A system having cognitive computing power should posses following three qualities (i) Sense, (ii) Comprehend, and (iii) Act. Sensing deals with the power of *"Computer VIsion"*. Comprehension signifies the role of *"NLP"* concept and *"Knowledge-Base"* in the system. Finally, Action describes the architectural way of system designing. Following *"Table 1"* describes the cognitive computing powers that *"SN : CQA"* posses. Rest techniques will be implemented soon.

Table 1. Comparison table for showing cognitive computing capabilities of "SN : CQA".

Cognitive property	Acheived	Implementing as future work
Sense:		
Speech Processing:	- -	Will be implemented
Audio & Visual Capability:	- -	Will be implemented
Comprehend:		
NLP Powered:	Yes	- -
Knowledge-Base:	Yes	- -
Act:		
System Approach:	Yes	- -
Agent Based System:	Yes	- -
Expert System:	Yes	- -

5   Conclusion and Future Work

This paper proposes "SN : CQA" (A Semantic-Network based NLP powered Cognitive Question-Answering system). It offers system approach based architectural designing of the system. The present methods are either proposing new techniques or talking about the implementation of new technologies. However for the better results, one should follow the "*Architectural Way*" of system designing which can synergies all the power of cognitive computing at one place. "SN : CQA" has designed based on "*System Approach*", and for achieving the "*Autonomy*", it implements the concept of "*Agent-Based Computing*". In the proposed architecture we have emphasized on the "*Experiential Knowledge-Base*". Here, domain expert (a Teacher) fills the facts in the knowledge-base. This concept improves the accuracy of answers and overcomes with the problems present in the web mining based answer extraction approach. The present systems are using the web mining approach for knowledge extraction which suffers from accuracy issue of answers. Subsequently, We have also implemented the concept of *Liking Value*. It provides the feedback or helps the knowledge engineer in knowledge updating and in filling up the new facts. The low like value for a particular answer signifies less understanding or the high complexity level for the specific answer. Accordingly, next time while inserting new facts in the corpus a teacher will take care of this (He has to provide easy answers to the questions or has to make changes accordingly). Shortly we will try to provide *Computer Vision*, *Audio Processing* and other cognitive computing powers to the system.

Acknowledgments. I would like to thank Dr. M. B. Lal Sahab and Dr. P. S. Satsangi Sahab for his continuous inspirations and blessings.

References

1. Prajapati, A.P., Chaturvedi, D.K.: Semantic network based knowledge representation for cognitive decision making in teaching electrical motor concepts. In: 2017 International Conference on Computer, Communications and Electronics (Comptelix), pp. 147–162. IEEE (2017) https://doi.org/10.1109/COMPTELIX.2017.8003954

2. Prajapati, A.P., Chaturvedi, D.K.: Ontology based knowledge representation for cognitive decision making in teaching electrical motor concepts. In: Silhavy, R., Senkerik, R., Kominkova Oplatkova, Z., Prokopova, Z., Silhavy, P. (eds.) CSOC 2017. AISC, vol. 573, pp. 43–53. Springer, Cham (2017). https://doi.org/10.1007/978-3-319-57261-1_5

3. Hu, Z., Zhang, Z., Yang, H., Chen, Q., Zuo, D.: A deep learning approach for predicting the quality of online health expert question-answering services. J. Biomed. Inform. **71** (2017), https://doi.org/10.1016/j.specom.2017.05.001

4. Jaya Kumar, A., Schmidt, C., Köhler, J.: A knowledge graph based speech interface for question answering systems. Speech Commun. **92**, 1–12 (2017). https://doi.org/10.1016/j.specom.2017.05.001

5. Figueroa, A.: Automatically generating effective search queries directly from community question-answering questions for finding related questions. Expert Syst. Appl. **77**, 11–19 (2017). https://doi.org/10.1016/j.eswa.2017.01.041

6. Yue, C., Cao, H., Xiong, K., Cui, A., Qin, H., Li, M.: Enhanced question understanding with dynamic memory networks for textual question answering. Expert Syst. Appl. **80**, 39–45 (2017). https://doi.org/10.1016/j.eswa.2017.03.006

7. Peng, P., Zou, L., Qin, Z.: Answering top-K query combined keywords and structural queries on RDF graphs. Inf. Syst. **67**, 19–35 (2017). https://doi.org/10.1016/j.is.2017.03.002

8. Fu, H., et al.: ASELM: adaptive semi-supervised ELM with application in question subjectivity identification. Neurocomputing **207**, 599–609 (2016). https://doi.org/10.1016/j.neucom.2016.05.041

9. Stevens, J.S., Benz, A., Reuße, S., Klabunde, R.: Pragmatic question answering: a game-theoretic approach. Data Knowl. Eng. **106**, 52–69 (2016). https://doi.org/10.1016/j.datak.2016.06.002

10. Olteeanu, A.-M., Falomir, Z.: comRAT-C: a computational compound remote associates test solver based on language data and its comparison to human performance. Pattern Recognit. Lett. **67**, 81–90 (2015). https://doi.org/10.1016/j.patrec.2015.05.015

11. Momtazi, S., Klakow, D.: Bridging the vocabulary gap between questions and answer sentences. Inf. Process. Manag. **51**, 595–615 (2015). https://doi.org/10.1016/j.ipm.2015.04.005

12. Neves, M., Leser, U.: Question answering for biology. Methods **74**, 36–46 (2015). https://doi.org/10.1016/j.ymeth.2014.10.023

13. Shekarpour, S., Marx, E., Ngomo, N.A.-C., Auer, S.: SINA: semantic interpretation of user queries for question answering on interlinked data. Web Semant. Sci. Serv. Agents World Wide Web **30**, 39–51 (2015). https://doi.org/10.1016/j.websem.2014.06.002

14. Procaci, T.B., Siqueira, S.W.M., Braz, M.H.L.B., Vasconcelos de Andrade, L.C.: How to and people who can help to answer a question? Analyses of metrics and machine learning in online communities. Comput. Hum. Behav. **51**, 664–673 (2015). https://doi.org/10.1016/j.chb.2014.12.026

15. Hattori, L., D'Ambros, M., Lanza, M., Lungu, M.: Answering software evolution questions: an empirical evaluation. Inf. Softw. Technol. **55**, 755–775 (2013). https://doi.org/10.1016/j.infsof.2012.09.001
16. Heie, M.H., Whittaker, E.W.D., Furui, S.: A Question answering using statistical language modelling. Comput. Speech Lang. **26**, 193–209 (2012). https://doi.org/10.1016/j.csl.2011.11.001
17. Bataller, C., Harris, J.: Turning Cognitive Computing into Business Value. Today, 21 May 2015. https://www.accenture.com

Novel Wrapper-Based Feature Selection for Efficient Clinical Decision Support System

R. Vanaja[✉] and Saswati Mukherjee

Department of Information Science and Technology, College of Engineering
Guindy, Anna University, Chennai, Tamil Nadu, India
vanajagokul@gmail.com, msaswati@auist.net

Abstract. Although healthcare sector has evolved with several new computer technologies it requires effective and efficient analytical techniques to truly exploit the benefits. As the industry is time sensitive in nature, there is an absolute need to perform medical diagnosis accurately without compromising the minimal time constraint. As predictive analytics is justified to be a suitable methodology that can be applied in healthcare sector, the proposed work uses the machine learning approach in a prospective way in performing effective learning of medical data. The proposed work aims to build an efficient prediction model using two novel feature selection approaches based on variants of Particle Swarm Optimization (PSO) named Particle Swarm Optimization with Digital Pheromones (PSODP) and a combination of PSO and PSODP. The research performs diagnosis for diabetic, breast cancer and chronic kidney disease data using deep learning. The proposed work shows improvement in classification accuracy with minimal time requirement compared to existing feature selection and classification techniques.

Keywords: Predictive analytics · Feature selection · Deep learning

1 Introduction

Analytics is the systematic use of data and related business insights developed through applied analytical disciplines (e.g. statistical, contextual, quantitative, predictive, cognitive, other models) to derive fact-based decision making for planning, management, measurement and learning [3]. In recent days, research work is focusing on predictive analytics, especially in clinical settings attempting to optimize health and financial outcomes. Predictive analytics is proving to be an effective way for developing mathematical models, algorithms that make predictions by applying a wide variety of mathematical techniques to historical data. As the predictive analytical capabilities mature, healthcare organizations are concentrating toward associated techniques that utilize an understanding of the past to predict future activities [2] and model scenarios using simulation and forecasting. These advanced capabilities support enterprise analysis, clinical outcome analysis, and evidence-based medicine.

The vital steps in the predictive analytics are the preprocessing phase and the machine learning phase. While the preprocessing phase is concerned with handling missing data, removing redundancy and making the right data available to the machine

L. Akoglu et al. (Eds.): ICIIT 2018, CCIS 941, pp. 113–129, 2019.
https://doi.org/10.1007/978-981-13-3582-2_9

learning phase, the machine learning phase is concerned with training the data and generating patterns in order to perform accurate predictions in the future [6]. Feature selection plays a pivotal role in the preprocessing phase and hence in the whole system [9]. Feature selection is automatic selection of the most relevant features or attributes for a given problem. This research proposes two novel feature selection methods, one using Particle Swarm Optimization with Digital Pheromones (PSODP) [10] that perform better than the existing feature selection approaches [27], and the other feature selection using the combination of PSO and PSODP. Also the research contributes in developing an effective prediction model that can be used by the medical practitioners for accurate diagnosis.

The efficacy of the proposed methods have been established using popular datasets for the diagnosis of diseases like diabetes, breast cancer and chronic kidney disease by using deep learning approaches like Feed Forward Neural Network (FFNN), Back Propagation Neural Network (BPNN) and Recurrent Neural Network (RNN) for the classification task. An important contribution in the world of healthcare since the ability to accurately predict the risk of serious outcomes that would allow healthcare professionals to focus on appropriate measures to reduce the risk of adverse diseases. Machine learning in healthcare provides countless possibilities for hidden pattern investigation and hence can be used by physicians to perform diagnosis, prognosis and hence to provide an effective treatment for patients.

2 Literature Survey

This section briefly explains the survey carried with due consideration to the feature selection process in developing an effective clinical decision support system. With the creation of huge databases and the consequent requirements for good machine learning techniques new problems arise and novel approaches to feature selection are in demand [4, 14].

2.1 Variants of PSO in Feature Selection

PSO have been used for feature selection [23, 24] for many applications particularly in intrusion detection and in cybernetics [22]. In the recent past of the research, feature selection using several variants of PSO has been explored. Some of its variants like the hybrid PSO for feature selection [15], velocity bounded Boolean PSO for the diagnosis of diseases like liver and chronic kidney [7] has been surveyed. Other variants include PSO based hybrid feature selection for high dimensional classification [19], quantum behaved PSO for diagnosis of cancer disease [21] and multi-objective PSO based approach for cost-based feature selection in classification [25].

A variant of PSO uses digital pheromones for aiding communication within the swarm to improve the search efficiency and reliability [11]. Digital pheromones have been used in the automatic adaptive swarm management of Unmanned Aerial Vehicles (UAVs) [20] where the costs of human operators are greatly reduced. Sree Hari Rao and Naresh Kumar have proposed novel approaches for predicting risk factors of Atherosclerosis [17]. Prasan Kumar Sahoo et al. have designed a probabilistic data acquisition method for the cloud-based healthcare system [18]. With the emphasis on cost consciousness and cost efficiency, Nabil Alshurafa et al. has reported a pilot study to predict the level of outcome success in reducing the risk factors for Cardio Vascular Disease (CVD) [1]. Darcy A Davis et al. has proposed a collaborative assessment and recommendation engine for prospective and proactive healthcare using ICD disease codes [5].

2.2 Deep Learning for Training

As the proposed work focuses more on classification techniques based on the availability of the datasets, deep learning techniques are explored in the research for training and testing the medical data. Deep learning algorithms are one promising avenue of research [16] into the automated extraction of complex data representations (features) at high levels of abstraction. Deep learning solutions have yielded outstanding results in different machine learning applications including speech recognition, computer vision, and natural language processing. The distributed feature representation of deep learning approaches is very powerful compared to other learning representations. RNN uses diagnosis data of adult disease patients to predict their prognosis to cardio-vascular diseases [8].

3 Motivation for the Proposed Research

This section explains about a variant of PSO technique named PSODP that has been proposed by Kalivarapu and Winer in [11] for exploring the solution space efficiently when compared to basic PSO. However, PSODP has not been explored for feature selection. The concept behind the PSODP technique and their novelty in their usage for feature selection is explained which is a strong motivation for the proposed research.

3.1 PSO with Digital Pheromones (PSODP) - An Unexplored Option for Feature Selection

Digital Pheromones

Pheromones are chemical scents produced by insects to communicate with each other and serve as a stimulus to invoke behavioral responses from creatures of their own species. The stronger the pheromone, the more the insects are attracted to the path. A digital pheromone is analogous to an insect generated pheromone in that it can be used as a marker to determine whether or not an area of a design space is promising for further investigation [12].

Target Pheromone

With numerous digital pheromones placed within the design space, it is crucial to compute a single target digital pheromone for each particle of the swarm. The criteria for this computation are (a) a small magnitude of distance from the particle and (b) a high pheromone level. Therefore, in order to rank which digital pheromone has the most influence and attraction, a target pheromone attraction factor P' is computed as mentioned below and a particle choosing the target pheromone [12] is depicted in Fig. 1.

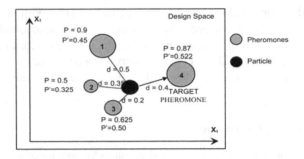

Fig. 1. A particle choosing its target pheromone [12]

$$P' = (1 - d)P$$

$$d = \sqrt{\sum_1^k \left(\frac{Xp_k - X_k}{range_k}\right)^2}$$

$k = 1 : n \#$ of design variables

$Xp = $ Location of pheromone

$X = $ Location of particle

In basic PSO, P number of particles is randomly distributed in a problem solution space S with N number of dimensions represented as S^N. Each particle will compute the solutions and determine their suitability by using the fitness function $f(s^1, s^2, \ldots, s^n)$, where $0 < n \le N$ and $s^n \in S^N$ [12]. The objective of the optimization is to find a set of $S' \subset S$ to maximize/minimize the fitness function $s' = f(s^1, s^2, \ldots, s^n)$. The PSO technique initially generates random positions and velocities for a population of particles [13]. Each particle represents an alternative solution in the multidimensional search space. Each particle computes its way through the search space with the velocity constantly updated according to its own search experience and its neighbour's best experience. The position vector and velocity vector of the ith particle in the D-dimensional search space can be represented as $X_i = (x_{i1}, x_{i2}, \ldots, x_{iD})$ and $V_i = (v_{i1}, v_{i2}, \ldots, v_{iD})$ respectively. According to a predefined fitness function mentioned as before, the best previous position of the ith particle among all the particles found so far is $P_g = (P_{g1}, P_{g2}, \ldots, P_{gD})$. The velocity and position of the particles are updated based on the Eqs. (1) and (2).

$$V_{id}(t+1) = w * v_{id}(t) + c1 * r1(t+1) * [P_{id}(t) - x_{id}(t)] + c2 \\ * r2(t+1)[P_{gd}(t) - x_{id}(t)] \tag{1}$$

$$x_{id}(t+1) = x_{id}(t) + V_{id}(t+1) \tag{2}$$

where t is the index of the iterations, w is the inertia weight, $c1$ and $c2$ are positive constants known as acceleration coefficients, $r1(t)$ and $r2(t)$ are two uniformly distributed random variables in the range (0,1) [11]. The second part of velocity update equation is known as cognitive component. It represents the personal thinking of each particle. This component encourages the particles to fly towards their own best positions found so far. The third part of velocity update shown in the above equation is the social component, which represents the co-operative effect of the particles in optimization searching. This component always leads the particles towards the global best particle found so far. Generally a maximum velocity vector V_{max} is defined and acts as an upper limit for the achievable velocity of the particles. It is used to control the ability of the particles to search and is often confined within the search space. The PSODP updates the velocity of particles based on Eq. (3), where a fourth component that uses target pheromone concept as explained in Fig. 1 is added.

$$V_{id}(t+1) = w * v_{id}(t) + c1 * r1(t+1) * [P_{id}(t) - x_{id}(t)] + c2 * r2(t+1) * [P_{gd}(t) - x_{id}(t)]$$
$$+ c3 * r3(t+1) * [\text{Target pheromone}_{id}(t) - x_{id}(t)]$$
(3)

3.2 Comparison Between PSO Search and PSODP Search

In a basic PSO algorithm, the swarm movement is governed by the velocity vector computed in Eq. (1). Each swarm member uses information from its previous best and the best member in the entire swarm at any iteration. However, multiple pheromones released by the swarm members could provide more information on promising locations within the design space when the information obtained from pBest and gBest (P_{id} and P_{gd} in Eq. 3) are insufficient or inefficient [10]. Figure 2a displays a scenario of a swarm member's movement whose direction is guided by pBest and gBest alone. If $c1 \gg c2$, the particle is attracted primarily towards its personal best position. On the other hand, if $c2 \gg c1$, the particle is strongly attracted to the gBest position.

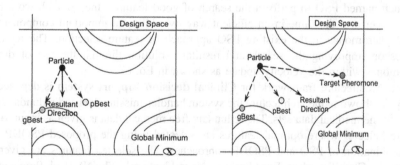

Fig. 2. a. Particle movement in Basic PSO **b.** Particle movement with digital pheromones [12]

Figure 2b shows the effect of implementing digital pheromones into the velocity vector. An additional pheromone component potentially causes the swarm member to result in a direction different from the combined influence of pBest and gBest thereby increasing the probability of finding the global optimum.

3.3 Feature Selection Using PSODP and Existing Approaches - A Comparison

On comparing the existing feature selection approaches and the proposed methods, the following are observed,

a. Most of the existing feature selection methods that use statistical approaches follow sequential processing and hence the above motivates the proposed work to be carried out using wrapper-based search approach using inbuilt parallelization method like PSODP.
b. While many of the feature selection that uses statistical methods perform well in terms of execution time, they underperform when accuracy aspect is considered. This motivates the research work to propose a method that performs both in terms of accuracy and execution time.
c. An appropriate combination of feature selection using PSODP, PSO+PSODP and classification using various deep learning approaches like Feed Forward Neural Network (FFNN), Back Propagation Neural Network (BPNN) and Recurrent Neural Network (RNN) develops a good prediction model applicable in clinical decision support system.

4 The Proposed Research

The proposed work mainly concentrates in developing an effective search strategy for wrapper based feature selection method that produces the optimal number of good features in order to improve the classifier performance in terms of accuracy. As the recent research in feature selection mostly uses the population based heuristic search approach named PSO to perform the search of good features, the research work proposes to perform the same in an efficient way by adding an important component of digital pheromones in the existing PSO approach of feature selection. The research focuses on improving the search and resultant solution through the use of digital pheromones within the velocity update as shown in Eq. (3).

The Architecture framework for **Clinical decision support system** is depicted in Fig. 3. As shown in the architecture the system handles missing data and redundancy of the real-time medical data sets. The redundant free medical data is given as input to the Feature Selector where optimal features are selected using the proposed PSODP and the combination of PSO and PSODP approaches. The selected features are given as input to the Classifier where Feed Forward Neural Network (FFNN), Back Propagation Neural Network (BPNN) and Recurrent Neural Network (RNN) are used to construct the necessary patterns and perform the classification. A new patient can determine the existence of a disease through the Predictor component on going through the patterns constructed during the learning process. The following sub-section explains in detail the important component of the architecture diagram.

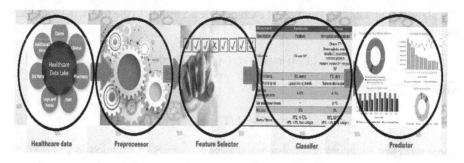

Healthcare data   Preprocessor   Feature Selector   Classifier   Predictor

Fig. 3. Architecture diagram of Clinical decision support system

4.1 Feature Selection Using Particle Swarm Optimization with Digital Pheromones (FS-PSO-DP)

The basic PSO method could cause the swarm to get locked into local optimum and if not, it takes very long time to reach the global optimum. The increased solution accuracy and the decreased solution time of using the digital pheromones concept in the basic PSO motivates to implement the proposed FS-PSO-DP approach for feature selection. The algorithm uses the binary implementation of FS-PSO-DP.

FS-PSO-DP algorithm.

Begin

> divide *Dataset* into a Training set and a Test set;
> randomly initialize the position and velocity of each particle;
> *while maximumiterations or the stopping criterion is not met do*
>> evaluate fitness of each particle according to Equation 6 and 7 ;
>> *for i=1 to populationsize do*
>>> update the *pbest* of particle *i*;
>>> update the *gbest* of particle *i*;
>>> update the *Targetpheromone* of particle *i*;
>>
>> *for i=1 to populationsize do*
>>> *for d=1 to numberofavailablefeatures do*
>>>> update the velocity of particle *i* according to Equation 3;
>>>> update the position of particle *i* according to Equations 4 and 5;
>
> calculate the classification accuracy of the selected feature subset on the test set;
> return the position of *gbest* (the selected feature subset);
> return the training and test classification accuracies;

The velocity here represents the probability of an element in the position taking value 1. Equation (3) is still applied to update the velocity while x_{id}, p_{id}, p_{gd} and *Targetpheromone$_{id}$* are restricted to 1 or 0. A sigmoid function $s(v_{id})$ is introduced to transform the obtained velocity values to the range of (0, 1) as mentioned in Eq. (5).

The algorithm updates the position of each particle according to the following formulae mentioned in the Eq. (4).

$$x_{id} = \{1, \text{ if rand}() < s(v_{id}) \text{ and } 0 \text{ otherwise} \tag{4}$$

where,

$$s(v_{id}) = 1/1 + e^{-Vid} \tag{5}$$

where rand() is a random number selected from a uniform distribution in [0,1].

Fitness Value Computation in FS-PSO-DP
The fitness function is to minimize the classification error rate (maximize the classification accuracy) obtained by the selected feature subset.

$$\text{Fitness} = \min(\text{Error\_rate}) \tag{6}$$

$$\text{Error\_rate} = (FP + FN)/(TP + TN + FP + FN) \tag{7}$$

where TP, TN, FP and FN are True Positive, True Negative, False Positive and False Negative values based on the test data. The representation of a particle in binary FS-PSO-DP is a n-bit binary string, where n is the number of available features in the dataset and also the dimensionality of the search space. In the binary string, "1" represents that the feature is selected and "0" otherwise.

4.2   Contribution of the Proposed Work

(a) The proposed work aims to implement the wrapper-based feature selection using the search strategy of PSODP individually. This implementation uses the Eq. 3 to update the velocities of the particles.

(b) The contribution also extends to implement the wrapper-based feature selection using the combination of PSO and PSODP whenever the condition of one or more particles cannot proceed beyond their local optimum arise in PSO. Here PSODP approach plays a vital role to bring the particles come out of their local optima and to explore the search space in an efficient manner as explained in Figs. 4 and 5.

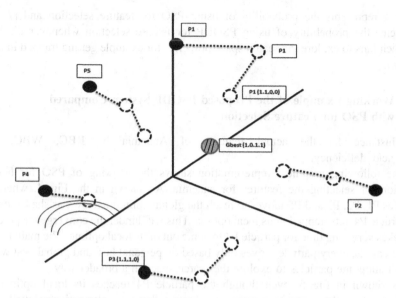

Fig. 4. Movement of particles based on PSO for feature selection

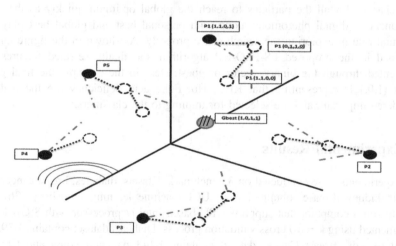

Fig. 5. Movement of particles based on PSODP for feature selection

4.3 The Proposed Velocity Update

The following velocity update is proposed in order to support the combination of PSO and PSODP approaches for feature selection.

$$
\begin{aligned}
V_{id}(t+1) = {} & \alpha(w * v_{id}(t) + c1 * r1(t+1) * [P_{id}(t) - x_{id}(t)] + c2 * r2(t+1) [P_{gd}(t) - x_{id}(t)]) \\
& + (1 - \alpha)w * v_{id}(t) + c1 * r1(t+1) * [P_{id}(t) - x_{id}(t)] + c2 * r2(t+1) [P_{gd}(t) - x_{id}(t)] \\
& + c3 * r3(t + 1) * [Target\,pheromone_{id}(t) - x_{id}(t)])
\end{aligned} \tag{8}
$$

where α represents the probability of using PSO for feature selection and $(1 - \alpha)$ represents the probability of using PSODP for feature selection whenever the PSO approach fails to explore the search space properly (for example getting trapped in local optima).

4.4 Working Example of the Proposed PSODP System Compared with PSO for Feature Selection

For Instance let the actual features of Anaemia be RBC, WBC, Hb, Folic_acid_deficiency.

The following graphical representation shows the working of PSO for feature selection for selecting the features for anaemia. As shown in the Fig. 4, when the particles P1, P2, P3 and P5 moves to reach the global optimum shown as shaded circle, the particle P4 gets trapped in its local optima. This will hinder the convergence process, as it takes very long time for particle P4 to come out of its local optima. The main reason for the case is, every particle moves only based on personal best and global best which doesn't allow the particles to explore the search space in a broader way.

As shown in Fig. 5, even though the particle P4 reaches its local optima, on realizing the pheromone component shown through alternative dotted line (i.e. $- \cdot \; - \cdot \; - \cdot$) it by-passes the local optima very easily and quickly. Now there are good chances for all the particles to reach the global optimum quickly as the third component of digital pheromone along with personal best and global best plays an important role in exploring the search space properly. As shown in the figure and as explained in the proposed FS-PSO-DP algorithm, the finally selected features are represented through the binary vector of gbest. Here in the example the final gbest shows (1,0,1,1) representing that RBC, Hb, Folic_acid_deficiency are the features considered important and are selected for training to the classifiers.

5 Experimental Results

The experiments were conducted on 3 benchmark datasets, diabetes, breast cancer and chronic kidney disease, obtained from UCI machine learning repository [26] and executed on a computer that supports CPU-based Intel i7 processor with 8 GB RAM implemented using k = 10 cross validation process. Diabetes dataset contains 1,39,761 patient records, Breast Cancer dataset contains 1,143 patient records and Chronic kidney disease contains 2,138 records. Various existing feature selection methods and classifiers namely Naïve Bayes, Decision Tree, Random Forest and SVM are implemented in order to compare with the proposed research work. The following tables show the analysis of the above datasets with respect to the various metrics of classification that are used for medical diagnosis. The performance of existing classifiers and classification using deep learning approaches (FFNN, BPNN, RNN) with various feature selection methods including the two proposed feature selection methods (PSODP and PSO+PSODP) have been analyzed (Table 1).

Table 1. Performance evaluation for diagnosis of diabetes

Feature selection methods	Classifiers used	Diabetes (Original features: 8, Features selected: 6)			
		Error rate	Accuracy	F1 Score	Time comp.
Pair-wise correlation	Naïve Bayes	30.050	69.950	0.7531	10 m 42 s
	Decision Tree	28.445	71.555	0.7738	9 m 8 s
	Random Forest	28.021	71.979	0.7811	10 m 18 s
	SVM	27.650	72.350	0.7925	10 m 23 s
	DL-FFNN	29.551	70.449	0.7934	11 m 40 s
	DL-BPNN	22.476	77.524	0.8379	11 m 6 s
	DL-RNN	21.580	78.420	0.8464	9 m 7 s
Eigen-Centrality	Naïve Bayes	30.955	69.045	0.7612	10 m 48 s
	Decision Tree	29.260	70.740	0.7731	9 m 18 s
	Random Forest	29.001	70.999	0.7800	9 m 58 s
	SVM	28.887	71.113	0.7882	10 m 3 s
	DL-FFNN	22.467	77.533	0.8319	11 m 50 s
	DL-BPNN	22.281	77.719	0.8394	10 m 40 s
	DL-RNN	23.011	76.989	0.8318	10 m 40 s
Latent	Naïve Bayes	31.417	68.583	0.7529	11 m 4 s
	Decision Tree	30.479	69.521	0.7650	9 m 28 s
	Random Forest	29.026	70.974	0.7713	10 m 8 s
	SVM	28.564	71.436	0.7821	10 m 46 s
	DL-FFNN	22.624	77.376	0.8367	11 m 19 s
	DL-BPNN	22.324	77.676	0.8367	10 m 50 s
	DL-RNN	24.156	75.844	0.8241	11 m 3 s
Relief	Naïve Bayes	30.159	69.841	0.7644	10 m 4 s
	Decision Tree	25.890	74.110	0.8012	9 m 18 s
	Random Forest	26.004	73.996	0.8001	10 m 25 s
	SVM	27.357	72.643	0.7986	10 m 3 s
	DL-FFNN	22.281	77.719	0.8393	10 m 4 s
	DL-BPNN	20.800	79.200	0.8611	9 m 4 s
	DL-RNN	20.223	79.777	0.8711	8 m 54 s
PSO	Naïve Bayes	30.159	69.841	0.7644	10 m 4 s
	Decision Tree	25.890	74.110	0.8012	9 m 18 s
	Random Forest	26.004	73.996	0.8001	10 m 25 s
	SVM	27.357	72.643	0.7986	10 m 3 s
	DL-FFNN	22.281	77.719	0.8393	10 m 4 s
	DL-BPNN	20.800	79.200	0.8611	9 m 4 s
	DL-RNN	20.223	79.777	0.8711	8 m 54 s
PSODP	Naïve Bayes	29.138	70.862	0.7712	7 m 58 s
	Decision Tree	24.075	75.925	0.8241	7 m 16 s
	Random Forest	24.221	75.779	0.8299	9 m 8 s

(*continued*)

Table 1. (*continued*)

Feature selection methods	Classifiers used	Diabetes (Original features: 8, Features selected: 6)			
		Error rate	Accuracy	F1 Score	Time comp.
	SVM	24.970	75.030	0.8246	8 m 59 s
	DL-FFNN	20.009	79.991	0.8541	9 m 30 s
	DL-BPNN	18.357	81.643	0.8898	7 m 20 s
	DL-RNN	14.432	85.568	0.9298	6 m 37 s
PSO+PSODP	**Naïve Bayes**	29.571	70.429	0.7795	7 m 8 s
	Decision Tree	24.999	75.001	0.8252	6 m 1 s
	Random Forest	24.351	75.649	0.8289	9 m 48 s
	SVM	25.012	74.988	0.8297	9 m 3 s
	DL-FFNN	20.008	79.072	0.8503	8 m 58 s
	DL-BPNN	19.275	80.725	0.8793	7 m 9 s
	DL-RNN	14.853	85.147	0.9266	5 m 4 s

From the above obtained results, it is seen that the proposed PSODP approach of feature selection performs good in terms of accuracy when it works with all classifiers and the proposed PSO+PSODP approach performs good in reducing the time complexity (Table 2).

Table 2. Performance evaluation for diagnosis of breast cancer

Feature selection methods	Classifiers used	Breast cancer (Original features: 9, Features selected: 7)			
		Error rate	Accuracy	F1 Score	Time comp.
Pair-wise correlation	**Naïve Bayes**	19.955	80.045	0.8143	6 m 32 s
	Decision Tree	17.835	82.165	0.8368	5 m 3 s
	Random Forest	17.998	82.002	0.8391	5 m 54 s
	SVM	16.154	83.846	0.8466	5 m 12 s
	DL-FFNN	11.754	88.246	0.8873	3 m 18 s
	DL-BPNN	1.754	98.246	0.9885	4 m 47 s
	DL-RNN	3.508	96.491	0.9750	1 m 48 s
Eigen-Centrality	**Naïve Bayes**	19.821	80.179	0.8163	6 m 44 s
	Decision Tree	18.998	81.002	0.8295	5 m 13 s
	Random Forest	18.108	81.892	0.8355	5 m 6 s
	SVM	19.002	80.998	0.8211	5 m 49 s
	DL-FFNN	11.754	88.246	0.8890	3 m 15 s
	DL-BPNN	1.500	98.500	0.9950	5 m 8 s
	DL-RNN	7.017	92.987	0.9556	1 m 6 s

(*continued*)

Table 2. (*continued*)

Feature selection methods	Classifiers used	Breast cancer (Original features: 9, Features selected: 7)			
		Error rate	Accuracy	F1 Score	Time comp.
Latent	**Naïve Bayes**	20.028	79.972	0.8055	6 m 8 s
	Decision Tree	19.959	80.041	0.8120	5 m 52 s
	Random Forest	18.997	81.333	0.8267	5 m 4 s
	SVM	17.913	82.087	0.8299	5 m 2 s
	DL-FFNN	11.5	88.5	0.8900	3 m 16 s
	DL-BPNN	3.508	96.491	0.9767	5 m 36 s
	DL-RNN	5.125	94.875	0.9875	1 m 11 s
Relief	**Naïve Bayes**	19.025	80.975	0.8264	6 m 48 s
	Decision Tree	18.717	81.283	0.8391	6 m 6 s
	Random Forest	18.024	81.976	0.8439	5 m 32 s
	SVM	18.500	81.500	0.8268	5 m 9 s
	DL-FFNN	11.75	88.246	0.8832	3 m 17 s
	DL-BPNN	1.750	98.250	0.9900	5 m 39 s
	DL-RNN	1.754	98.246	0.9912	1 m 7 s
PSO	**Naïve Bayes**	18.778	81.222	0.8340	4 m 24 s
	Decision Tree	16.328	83.672	0.8539	3 m 7 s
	Random Forest	15.157	84.843	0.8691	3 m 3 s
	SVM	15.887	84.113	0.8622	3 m 54 s
	DL-FFNN	10.257	89.743	0.9026	1 m 36 s
	DL-BPNN	1.500	98.500	0.9950	1 m 13 s
	DL-RNN	1.500	98.500	0.9936	1 m
PSODP	**Naïve Bayes**	17.332	82.668	0.8561	3 m 22 s
	Decision Tree	15.856	84.144	0.8672	2 m 58 s
	Random Forest	15.002	84.998	0.8731	2 m 41 s
	SVM	14.992	85.008	0.8842	2 m 8 s
	DL-FFNN	9.85	90.15	0.9312	1 m 3 s
	DL-BPNN	0.750	99.250	0.9999	58 s
	DL-RNN	0.569	99.431	0.9998	50 s
PSO+PSODP	**Naïve Bayes**	17.385	82.615	0.8539	3 m 50 s
	Decision Tree	15.916	84.084	0.8820	2 m 25 s
	Random Forest	14.776	85.224	0.8922	2 m 10 s
	SVM	14.088	85.912	0.8996	2 m 40 s
	DL-FFNN	9.50	90.50	0.9398	58 s
	DL-BPNN	0.832	99.168	0.9987	1 m
	DL-RNN	0.5	99.5	0.9999	42 s

On evaluating the performance measures for the diagnosis of Breast cancer dataset, it is observed from the results that both the proposed feature selection methods performs good in terms of accuracy and time complexity (Table 3).

Table 3. Performance evaluation for diagnosis of chronic kidney

Feature selection methods	Classifiers used	Chronic kidney (Original features: 25, Features Selected: 18)			
		Error rate	Accuracy	F1 Score	Time comp.
Pair-wise correlation	**Naïve Bayes**	21.669	78.331	0.8096	12 m 22 s
	Decision Tree	19.471	80.529	0.8433	11 m 4 s
	Random Forest	18.998	81.002	0.8564	11 m 8 s
	SVM	19.024	80.976	0.8511	11 m 9 s
	DL-FFNN	5.789	94.211	0.9638	13 m 38 s
	DL-BPNN	4.358	95.642	0.9699	14 m 42 s
	DL-RNN	3.871	96.129	0.9788	9 m 24 s
Eigen-Centrality	**Naïve Bayes**	22.899	77.101	0.7953	12 m 59 s
	Decision Tree	20.569	79.431	0.8177	11 m 43 s
	Random Forest	19.995	80.005	0.8366	11 m 38 s
	SVM	19.416	80.584	0.8419	11 m 6 s
	DL-FFNN	7.9	92.100	0.9412	13 m 8 s
	DL-BPNN	6.289	93.711	0.9577	14 m 41 s
	DL-RNN	4.996	95.004	0.9733	9 m 32 s
Latent	**Naïve Bayes**	22.199	77.801	0.8076	12 m 19 s
	Decision Tree	20.726	79.274	0.8421	11 m 23 s
	Random Forest	19.879	80.121	0.8577	11 m 37 s
	SVM	19.022	80.978	0.8678	11 m 8 s
	DL-FFNN	8.045	91.955	0.9410	13 m 28 s
	DL-BPNN	6.297	93.703	0.9599	14 m 5 s
	DL-RNN	4.113	95.887	0.9799	9 m 10 s
ReliefF	**Naïve Bayes**	22.500	77.500	0.8011	12 m 9 s
	Decision Tree	21.714	78.286	0.8834	11 m 3 s
	Random Forest	19.900	80.100	0.8458	11 m 3 s
	SVM	19.207	80.793	0.8499	11 m 12 s
	DL-FFNN	9.015	90.985	0.9359	13 m 28 s
	DL-BPNN	7.734	92.266	0.9522	14 m 5 s
	DL-RNN	4.689	95.311	0.9861	9 m 18 s
PSO	**Naïve Bayes**	17.554	82.446	0.8561	12 m 29 s
	Decision Tree	14.238	85.762	0.8838	11 m 13 s
	Random Forest	14.887	85.113	0.8810	11 m 32 s
	SVM	14.837	85.163	0.8813	11 m 24 s
	DL-FFNN	7.857	92.143	0.9584	11 m 28 s
	DL-BPNN	3.500	96.500	0.9806	12 m 5 s
	DL-RNN	2.478	97.522	0.9901	8 m 18 s

(continued)

Table 3. (*continued*)

Feature selection methods	Classifiers used	Chronic kidney (Original features: 25, Features Selected: 18)			
		Error rate	Accuracy	F1 Score	Time comp.
PSODP	**Naïve Bayes**	17.356	82.644	0.8533	10 m 32 s
	Decision Tree	14.995	85.005	0.8802	10 m 30 s
	Random Forest	14.067	85.933	0.8893	10 m 45 s
	SVM	14.278	85.722	0.8865	10 m 12 s
	DL-FFNN	5.857	94.143	0.9701	9 m 48 s
	DL-BPNN	3.115	96.885	0.9892	9 m 25 s
	DL-RNN	1.677	98.323	0.9945	6 m 41 s
PSO+PSODP	**Naïve Bayes**	17.002	82.998	0.8475	9 m 42 s
	Decision Tree	15.648	84.652	0.8623	8 m 22 s
	Random Forest	14.002	85.998	0.8841	8 m 15 s
	SVM	13.879	86.121	0.8898	8 m 42 s
	DL-FFNN	4.550	95.450	0.9712	3 m 41 s
	DL-BPNN	1.382	98.618	0.9914	1 m 58 s
	DL-RNN	0.500	99.5	0.9986	1 m

With an example for high-dimensional dataset, the observed experimental values of chronic kidney disease shows that the second proposed PSO+PSODP approach outperforms other existing feature selection techniques in reducing the time complexity drastically.

6 Conclusion and Future Work

The proposed research has implemented two novel feature selection approaches based on evolutionary computational method that uses a variant of PSO. Deep learning techniques like FFNN, BPNN and RNN have been used for classification. The proposed variant namely PSODP and the combined PSO+PSODP has performed better compared to the existing filter-based statistical feature selection approaches and wrapper based PSO approach in terms of achieving good classifier accuracy. Traditionally when there is a trade-off between the time complexity and the classifier accuracy, the prediction model on using the proposed approaches has given a very closely accurate diagnosis of diseases in possible minimal time without compromising the classifier accuracy. As a part of future enhancement of the proposed work, apart from considering the important features into account, the characteristics of the features and its impact in classification also may be considered in designing a detailed analytics of other real time open source datasets.

Acknowledgement. We would like to thank and acknowledge the "Visvesvaraya PhD Scheme", MeitY, New Delhi for supporting the research financially in the form of scholarship.

References

1. Alshurafa, N., Sideris, C., Pourhomayoun, M., Kalantarian, H., Sarrafzadeh, M., Eastwood, J.A.: Remote health monitoring outcome success prediction using baseline and first month intervention data. IEEE J. Biomed. Health Inform. **21**(2), 507–514 (2017)
2. Cohen, I.G., Amarasingham, R., Shah, A., Xie, B., Lo, B.: The legal and ethical concerns that arise from using complex predictive analytics in health care. Health Aff. **33**(7), 1139–1147 (2014)
3. Cortada, J.W., Gordon, D., Lenihan, B.: The value of analytics in healthcare. IBM Institute for Business Value IBM, Global Business Service (2012)
4. Dash, M., Liu, H.: Feature selection for classification. Intell. Data Anal. **1**(3), 131–156 (1997)
5. Davis, D.A., Chawla, N.V., Blumm, N., Christakis, N., Barabasi, A.L.: Predicting individual disease risk based on medical history. In: Proceedings of the 17th ACM Conference on Information and Knowledge Management, pp. 769–778. ACM, October 2008
6. Fatima, M., Pasha, M.: Survey of machine learning algorithms for disease diagnostic. J. Intell. Learn. Syst. Appl. **9**(1), 1 (2017)
7. Gunasundari, S., Janakiraman, S., Meenambal, S.: Velocity bounded boolean particle swarm optimization for improved feature selection in liver and kidney disease diagnosis. Expert Syst. Appl. **56**, 28–47 (2016)
8. Ha, J.W., et al.: Predicting high-risk prognosis from diagnostic histories of adult disease patients via deep recurrent neural networks. In: 2017 IEEE International Conference on Big Data and Smart Computing (BigComp), pp. 394–399. IEEE, February 2017
9. https://www.analyticsvidhya.com/blog/2016/12/introduction-to-feature-selection-methods-with-an-example-or-how-to-select-the-right-variables/
10. Kalivarapu, V.K.: Improving solution characteristics of particle swarm optimization through the use of digital pheromones, parallelization, and graphical processing units (GPUs) (2008)
11. Kalivarapu, V., Winer, E.: A statistical analysis of particle swarm optimization with and without digital pheromones. In: 48th AIAA/ASME/ASCE/AHS/ASC Structures, Structural Dynamics, and Materials Conference, p. 1882, April 2007
12. Kalivarapu, V., Foo, J.L., Winer, E.: Improving solution characteristics of particle swarm optimization using digital pheromones. Struct. Multidiscip. Optim. **37**(4), 415–427 (2009)
13. Kennedy, J.: Particle swarm optimization. In: Encyclopedia of Machine Learning, pp. 760–766. Springer, Boston (2011)
14. Kumar, V., Minz, S.: Feature selection. SmartCR **4**(3), 211–229 (2014)
15. Moradi, P., Gholampour, M.: A hybrid particle swarm optimization for feature subset selection by integrating a novel local search strategy. Appl. Soft Comput. **43**, 117–130 (2016)
16. Najafabadi, M.M., Villanustre, F., Khoshgoftaar, T.M., Seliya, N., Wald, R., Muharemagic, E.: Deep learning applications and challenges in big data analytics. J. Big Data **2**(1), 1 (2015)
17. Rao, V., Naresh Kumar, M.: Novel approaches for predicting risk factors of atherosclerosis (2015). arXiv preprint: arXiv:1501.07093
18. Sahoo, P.K., Mohapatra, S.K., Wu, S.L.: Analyzing healthcare big data with prediction for future health condition. IEEE Access **4**, 9786–9799 (2016)
19. Tran, B., Zhang, M., Xue, B.: A PSO based hybrid feature selection algorithm for high-dimensional classification. In: 2016 IEEE Congress on Evolutionary Computation (CEC), pp. 3801–3808. IEEE, July 2016
20. Walter, B., Sannier, A., Reiners, D., Oliver, J.: UAV swarm control: calculating digital pheromone fields with the GPU. J. Def. Model. Simul. **3**(3), 167–176 (2006)

21. Xi, M., Sun, J., Liu, L., Fan, F., Wu, X.: Cancer feature selection and classification using a binary quantum-behaved particle swarm optimization and support vector machine. Comput. Math. Methods Med. **2016**, 1–9 (2016)
22. Xue, B., Zhang, M., Browne, W.N.: Particle swarm optimization for feature selection in classification: a multi-objective approach. IEEE Trans. Cybern. **43**(6), 1656–1671 (2013)
23. Xue, B., Zhang, M., Browne, W.N.: Particle swarm optimisation for feature selection in classification: novel initialisation and updating mechanisms. Appl. Soft Comput. **18**, 261–276 (2014)
24. Xue, B., Zhang, M., Browne, W.N., Yao, X.: A survey on evolutionary computation approaches to feature selection. IEEE Trans. Evol. Comput. **20**(4), 606–626 (2016)
25. Zhang, Y., Gong, D.W., Cheng, J.: Multi-objective particle swarm optimization approach for cost-based feature selection in classification. IEEE/ACM Trans. Comput. Biol. Bioinform. (TCBB) **14**(1), 64–75 (2017)
26. https://archive.ics.uci.edu/ml/datasets.html
27. Vanaja, R., Saswati, M.: An effective clinical decision support system using swarm intelligence. Paper accepted and Abstract published in the proceedings of 1st International Symposium on Artificial Intelligence and Computer Vision, Chennai, Tamilnadu, India (2018). ISBN 978-93-84389-18-5

Linear and Nonlinear Analysis of Cardiac and Diabetic Subjects

Ulka Shirole[1]([✉]), Manjusha Joshi[2], and Pritish Bagul[3]

[1] Electronics Engineering Department, A.C. Patil College of Engineering,
Navi Mumbai, India
umshirole@acpce.ac.in
[2] Electronics and Telecommunication Department, MPSTME,
NMIMS Deemed to be University, Mumbai, India
manjusha.joshi@nmims.edu
[3] B and J super specialty hospital, Kamothe, Navi Mumbai, India
pritish.bagul@bjhospital.org

Abstract. Cardiac health issues are severe and cause maximum death according to the survey done by "World Health Organization" (WHO). Cardiac diseases are caused due to family history, living style, diabetes, etc. Diagnosis of cardiac health prior to pathological conditions is highly important. Heart Rate Variability (HRV) is the technique used to study the cardiac abnormalities, which are related to fluctuation in the sympathetic and parasympathetic activities. In this paper, we compare the time domain, frequency domain and nonlinear parameters of heart rate variability for 73 subjects. Our results show that HRV parameters are high for normal subjects compared to diabetic subjects and lowest for cardiac subjects. The results are validated by diagnosis done through clinical processes. Thus non-invasive ECG and HRV techniques help to diagnose the subject before it causes the cardiac arrest.

1 Introduction

Cardiac diseases include heart and blood vessel abnormalities leading to myocardial infarction (heart attack). These diseases are caused by various conditions such as poorly controlled high blood pressure, high cholesterol, diabetes, smoking, and family history. Due to these factors, heart supply insufficient oxygenated blood as per the needs of the body. Cardiac health issues are severe and cause maximum death according to the survey done by world health organization [1]. The time from the onset of cardiac diseases to the treatment for the same is very crucial. Immediate medical facility for such patient is essential before the patient's condition becomes beyond the control of the physician. The onset of cardiac diseases results in the changes in the ECG signal of the patient. These changes can be used as a tool for diagnosis. In such cases if the disease is confirmed severe the paramedical staff can start the preparatory treatment before the physician arrives. Various clinical devices and techniques such as 2D Echocardiogram, blood test, are available to analyze the heart's functionality [1]. Among all these techniques

L. Akoglu et al. (Eds.): ICIIT 2018, CCIS 941, pp. 130–140, 2019.
https://doi.org/10.1007/978-981-13-3582-2_10

quick confirmation is possible by using signal processing techniques applied to a normal ECG and the ECG of the patient under diagnosis [15].

ECG signal is a non-invasive method to understand the health of the heart and its functionality. Heart Rate Variability (HRV) is a technique used to study heart by analyzing the ECG signal. Heart rate variability is the degree of fluctuation in the length of the intervals between heartbeats. In other words, Heart Rate Variability is the variation in the beat to beat interval of the ECG signal. In clinical practices, doctors visually observe a short ECG waveform (2–7 s) to diagnose the heart functionality and very rarely HRV parameters are considered. However, to detect cardiac issues along the ECG waveform (5–15 min) and its analysis is important.

HRV is a useful tool for risk assessment in patients suffering from congestive heart failure. Acharya et al. [13], have developed a new novel methodology to automatically classify the HRV signals of normal and subjects at risk of sudden cardiac death by using nonlinear techniques. Nedim Soydam et al. [14], have shown that decreased HRV is associated with enhanced mortality due to the abnormal cardiac rhythm. While hypoglycemic events are increasingly common in the treatment of type 2 diabetes, HRV is part of the counter-regulation against low blood glucose levels.

Garciaa et al. [16], presented a new algorithm using RHRV package of HRV toolkit, which supports wavelet-based spectral analysis to perform HRV power spectrum analysis based on the Maximal Overlap Discrete Wavelet Packet Transform (MODWPT). The algorithm calculates power in any spectral band with a given tolerance for the band's boundaries. In his work, he developed the decomposition tree in required an optimum time by avoiding calculating unnecessary wavelet coefficients. The Paper shows how MODWPT based method overcomes the drawbacks of short time Fourier transform (STFT) and windowing burg method. Some quick changes missed in STFT and in windowed Burg method are successfully identified by the MODWPT. Their results suggest that among all windowing techniques wavelet-based technique analyze quick changes in RR time series [16].

Joshi and et al. Studied pathophysiological analysis of the correlation between Heart Rate Variability (HRV) parameter named Standard Deviation of NN interval (SDNN) and echocardiogram index called Left Ventricle Ejection Fraction (LVIF) for diabetic subjects having hypertension and without hypertension. The purpose of their research is to validate HRV analysis Results. Results show SDNN and LVEF indices values for normal cohort exceeds diseased cohorts. Further, the correlation between SDNN and LVEF is positive for diseased cohort, proving that SDNN is a good indicator for ECG analysis [17].

Cornforth et al. described in their work that classification of subjects by applying Renyi entropy gave higher accuracy than with time domain parameters alone. Also, Renyi entropy provided important information about time signal characteristic that may be helpful in clinical decision making [19].

Researchers have studied ECG signal with the help of HRV method using a number of analysis techniques such as linear [8,9], nonlinear [3,8], geometric

[2] etc. A linear system is defined by one or more linear equations [2]. Linear analysis of the ECG signal is done by time domain and frequency domain HRV parameters such as R-R interval, SDNN, PNN50, etc. A non-linear system is defined by the second order or higher order power system. Non-linear analysis of the ECG signal is done by entropy, Pioncare plot, etc.

Researchers have done an independent study of the cardiac and diabetic subjects. The correlation study of the diabetic and cardiac subjects is not done. In this paper, we have carried out a detailed linear and non-linear analysis of the cardiac and diabetic subjects. The paper is organized as follows: Sect. 2 describes the data collection, Sect. 3 presents analysis methods and results are illustrated in Sect. 4. Finally, Sect. 5 concludes the paper.

2   Data Collection

Clinical data is collected from 73 subjects, where 25 subjects were suffering from congestive heart failure, 23 subjects were diabetes and the remaining 25 were healthy subjects. Subjects involved in the study have age in the range of 30–65 years. ECG data were collected for these subjects from B and J super specialty hospital Kamothe, Navi Mumbai, India. Patients' ECG data is collected using a 12 electrode ECG acquisition machine based on the PC ECG software. PC-ECG software acquires data of all 12 channels and stores data separately for each channel in XML file format. ECG data of subjects is recorded for 15 min in the sitting position and 15 min in the supine position. The sampling rate of 500 Hz is used. In the PC ECG software, low pass filter is set for the cut-off frequency of 15 Hz to remove unwanted high-frequency components and a high pass filter is set for the cut-off frequency of 0.3 Hz to remove the low-frequency baseline wander.

2.1   Analysis Procedure

The complete procedure of analysis is described in Fig. 1. Digital data of cardiac, diabetic and normal subjects is collected using 12 electrode ECG machine supported by PC ECG software, that generates samples of ECG time series signal (Fig. 2) in XML form (Fig. 3) separately in a single file. For analysis and cardiac study samples from channel 2 is selected as it is near to the left ventricle. Special characters present in the data sequences are removed before applying it to Kubios software.

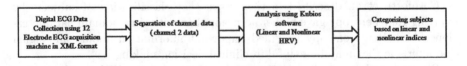

Fig. 1. Data analysis procedure

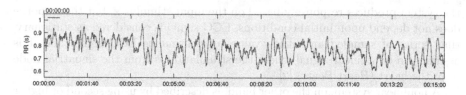

Fig. 2. RR-Interval time series

> .....-7 -15 -20 -27 -35 -41 -46 -50 -54 -57 -58 -59 -61 -64 -67 -67 -65 -67 -66 -66 -65 -67 -66 -64 -63 -62 -61 -61 -
> 61 -60 -58 -53 -53 -50 -49 -48 -47 -48 -48 -47 -47 -44 -40 -29 -15 6 37 77 122 170 226 286 343 402 461 514 556
> 587 606 613 609 596 571 531 482 430 375 318 259 206 157 113 79 51 33 18 12 12 14 18 22 25 30 31 34.....

Fig. 3. Sampled data from a single channel

Kubios software computes all linear time domain parameters like SDNN, NN50, frequency domain parameters like LF, HF, LF/HF, and also nonlinear parameters like Pioncare plot, entropy and so on. Software supports inputs in several formats [10].

It also supports trend removal, artifact correction, and QRS detection. The analysis results are stored in Matlab MAT- file form or PDF form. Linear and nonlinear indices were tabulated separately for normal, cardiac and diabetic subjects. Values of these indices are able to separate subject of one category from other.

3   Analysis Methods

ECG signal consists of different characteristics of heart which needs to be extracted during analysis using statistical methods. The objective of the analysis method is to extract the important hidden dynamical properties of the physiological phenomena. ECG signal analysis is done using time domain, frequency domain and nonlinear methods. HRV is the variation in the beat to beat interval of ECG signal. Different algorithms to process the HRV signals are available. These methodologies are broadly categorized into linear and nonlinear analysis. Linear analysis of the signals is done by using time domain analysis, frequency domain analysis and geometric analysis method, whereas for analyzing HRV signal using nonlinear method, the ideas like entropy, Poincare plot, and complexity are calculated [2].

A linear system is defined completely by one or more linear equations and has no exponent term. If a system generates values measured at different time intervals, the result consists of a time series consisting of discrete values. The linear numerical description of such series data consists of first power mathematical equation.

$$y = a + bx \tag{1}$$

the system produces response y based on the input stimulus x, but the response does not depend upon initial conditions. ECG signal is considered as stationary, time series of sampled amplitude values of continuous ECG signal and it is assumed to be a linear signal where the stimulation is from the sinuatrial node from the heart (Heart Beats) [2].

A nonlinear system, on the other hand, is described by using second or higher power of independent variable The simplest form of the nonlinear equation is

$$y = x^2 \tag{2}$$

In the nonlinear system, the independent variable contributes to the response. Nonlinear systems constituent parts cannot be analyzed separately as in linear system. The constituent parts in nonlinear system interfere, compete or cooperate with each other. With respect to the heart beats, ECG signal is influenced by the respiratory system, neural system and so on [2].

3.1   Linear Analysis

In this study, we consider only time domain and frequency domain methods. Time domain analysis is a study of variation of RR intervals with respect to time between successive cardiac cycles. It describes the autonomous nervous system (ANS) activity. Whereas, frequency domain analysis is the study of ECG signals in the various ranges of frequency. It describes ANS balance [8].

Time Domain Analysis - Time domain analysis includes the measurement of parameters like RR interval, standard deviation of NN interval (SDNN), standard deviation of the average NN interval (SDANN), heart rate (HR), percentage of number of successive NN interval which differ by more than 50 ms (pNN50) and root mean square root of successive difference (rMSSD). These parameters are calculated from variation in the intervals between successive cardiac cycles.

RR Interval: it is the time between two consecutive QRS complexes. The instantaneous heart rate can be calculated from the time between any two QRS complexes. It is measured in milliseconds. It is the degree of fluctuation in the length of intervals between heartbeats. It mirrors the regularity of heartbeats, i.e. greater the regularity lower the HRV and vice versa. The regularity of heartbeat is derived from a quantity of numbers, equal to the time elapsed between successive heart beats as shown in Fig. 4.

SDNN: It is the standard deviations of NN (RR) intervals. It is a global index of HRV. It is normally calculated using the standard deviation of normal QRS distances. It correlates strongly with total power. SDNN is the square root of NN interval variance. A variance is mathematically equivalent to the total power of

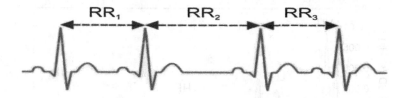

Fig. 4. R-R Interval of the cardiac signal

spectral analysis hence it reflects all cyclic components of variability in recorded series of NN intervals. It is calculated as follows:

$$SDNN = \sqrt{\frac{1}{N-1} \sum_{j=1}^{N} (RR_j - \overline{RR})^2} \qquad (3)$$

SDNN is calculated over a short period usually five minutes. SDNN is measured in milliseconds.

SDANN: Standard deviation of the average NN intervals calculated over short periods. Segments in the RR time series are calculated over short periods and their average is taken to give results. SDANN is calculated over short periods usually five minutes, therefore, requires longer measuring periods.

pNN50: NN50 is the number of pairs of successive NN intervals that differ by more than 50 ms. pNN50 is the proportion of NN50 divided by total number of NNs and it shows cardiac parasympathetic activity.

rmSSD: it is calculated through squaring of each NN interval this estimates high-frequency variations in heart rate. rmSSD estimates electrical stability and high-frequency variations in heart rate of short-term NN recordings. It reflects parasympathetic regulations of the heart. It is calculated by taking the square root of the mean squared differences of successive NN intervals. It is measured in milliseconds.

$$SDNN = \sqrt{\frac{1}{N-1} \sum_{j=1}^{N} (\varDelta RR_j)^2} \qquad (4)$$

Frequency Domain Analysis - Frequency domain analysis describes in principle the oscillations of the heart rate signal, decomposed at different frequencies and amplitudes, and provides information on the amount of their relative intensity in the sinus rhythm of the heart. The frequency domain is based on different frequency areas in the time series. The power spectral density is measured by calculating FFT (fast Fourier transform) of the time series. Typical variables include VLF (very low-frequency power), LF (low-frequency power) and HF (High-Frequency power) as shown in Fig. 5. A frequently used ratio is LF/HF, this variable is feasible both in short-term and long-term use.

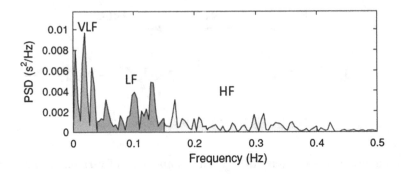

Fig. 5. FFT spectrum of ECG signal representing different frequency bands

VLF: Very Low-Frequency power band is between 0.003 Hz–0.05 Hz. In the short-term analysis, VLF does not provide much meaning since this band often reflect a meaningless noise signal. But VLF is associated with physical activity and possibly reflects sympathetic activity. It is not used as often but it is associated with clinical consequences than LF. Patients with the obstructive sleep apnea syndrome have been shown to have an increase in VLF band during the apnea, i.e., power in the VLF band increases when there is an absence of air exchange. Moreover, decreased VLF is associated with increased inflammatory parameters like CRP, I1–6, and WBC [2].

LF: Low-Frequency power band is between 0.05 Hz–0.15 Hz. Low-frequency component can reflect both sympathetic and parasympathetic activity. The very Low-Frequency band is calculated in milliseconds squared (ms2). Modulation in sympathetic activity (mental, physical stress, sympathomimetic pharmacological agents) results in an increase in LF power. A beta blockade leads to a decrease of LF. LF is not necessarily correlated with increased activity, but in the case of subjects with congestive heart failure patients, it is inversely correlated with sympathetic activity [2].

HF: High-Frequency power band is between 0.15 Hz–0.4 Hz. HF is dependent on respiration pattern, it is generally interpreted as a marker of parasympathetic modulation of ANS, it is partially associated with respiratory sinus arrhythmia.

LF/HF: this ratio reflects the global sympathovagal balance. It is feasible for both short-term and long-term use.

Non-linear Analysis. It is based on the fact that the ECG signal is non-stationary. The non-linear method can start without specifying any fixed pattern in the signal, but simply looking at similarities in the signal. In the non-linear analysis, Poincare plot and entropy are used [2].

Poincare Plot: Poincare plot analysis is a nonlinear geometrical method to assess the dynamics of the HRV. It is a diagram in which each R-R interval is plotted as a function of the previous R-R interval where the values of each pair of successive R-R interval define a point in the plot as shown in Fig. 6. It is the transform of time series data into phase series data. A quantitative analysis of HRV is displayed by the Poincare plot and can be made by adjusting to an ellipse. From the evaluation parameters like the area of an ellipse, the minor axis SD1 (standard deviation 1) and major axis SD2 (standard deviation 2) the performance analysis can be done. The ratio SD1/SD2 that are too low or too high are connected with illness. The diabetic and cardiac subjects have lower SD2 as compared to healthy subjects [4, 5].

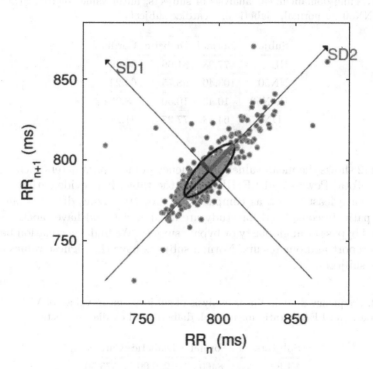

Fig. 6. Poincare plot of the subject indicating SD1 and SD2

Approximate Entropy: It is a measure of complexity. It quantifies the unpredictability of fluctuations in a time series [7]. This is a function for computing approximate entropy. To run it you need to provide the data stream, the window size and distance measure for comparison. A small value of entropy indicates regularity whereas higher numbers indicate the lower fraction of order or patterns in the data set.

4   Results and Discussion

Linear and non-linear analysis is carried out for collected data. Linear analysis of time domain parameters such as HR, SDNN, NN50, PNN50 are calculated and their mean values are shown in Table 1. From the Table 1, it is observed that the Mean HR value is high for Cardiac and diabetic patients; whereas for normal subjects, it is lowest among the three. Moreover, other parameters SDNN, NN50, pNN50 values are indicating a similar trend in reading, i.e., ascending scale from normal to the cardiac subject. These results show that the cardiac patients are at higher risk than that of diabetes.

Table 1. Time domain linear analysis of subjects, mean values of RR, SDNN, and NN50, pNN50 for normal, diabetic and cardiac subjects

Subjects	Normal	Diabetic	Cardiac
HR	77.08	81.66	80.46
NN50	103.40	98.75	64.22
PNN50	10.45	10.50	8.094
SDNN	64.13	77.27	41.28

Table 2 shows the mean values of frequency domain parameters such as VLF, LF. HF, Total Power, and LF/HF. From the table, it is evident that Cardiac subjects have least values as compared to subjects having diabetes only. The cardiac patients considered for study are at high risk and have blockages, and suggested bypass or angioplasty or bypass surgery. We find a correlation between doctor's report and our results. Normal subjects have the highest values among all three subjects.

Table 2. Frequency domain linear analysis of subjects, mean values of VLF, LF, HF, total power, and LF/HF ratio for normal, diabetic and cardiac subjects

Subjects	Normal	Diabetic	Cardiac
VLF	8460.20	1299.00	575.50
LF	2856.90	1874.00	646.83
HF	8581.70	3227.00	1088.39
Total power	49527.80	6425.00	2316.60
LF/HF	2.21	2.08	1.67

Table 3 shows a comparison between mean values of nonlinear parameters for normal, diabetic and cardiac subjects. Similar results are observed as in time domain and frequency domain, lowest values for cardiac and highest for normal as earlier proved by researchers.

Table 3. Non-linear analysis of subjects, mean values of SD1, SD2, SD1/SD2, and an entropy for normal, diabetic and cardiac subjects

Subjects	Normal	Diabetic	Cardiac
SD1	65.36	51.4	30.5
SD2	84.57	71.25	47.24
SD1/SD2	0.72	0.77	0.65
ApEn	1.21	1.01	1.16

5 Conclusion

Cardiac health assessment through ECG is non-invasive method is easy and quickly verifiable. However, detailed linear and non-linear analysis gives the confidence in the diagnosis of the patient. This paper describes the various time domain and frequency domain techniques to assess the cardiac health. Experimental results show that the analysis parameters are lowest for cardiac subjects than diabetic subjects as compared to healthy subjects.

References

1. http://www.who.int/mediacentre/factsheets/fs317/en/
2. Ernst, G.: History of heart rate variability. Heart Rate Variability, pp. 3–8. Springer, London (2014). https://doi.org/10.1007/978-1-4471-4309-3_1
3. Acharya, R., Faust, O., Kadri, N., Suri, J., Yu, W.: Automated identification of normal and diabetes heart rate signals using non-linear measures. Int. J. Comput. Biol. Med. (2013)
4. Golińska, A.: Poincare plots in analysis of selected biomedical signals. Stud. Logic Grammar Rhetoric **35**(1), 117–127. https://doi.org/10.2478/slgr-2013-0031
5. Joshi, M., Desai, K., Mennon, M.: Poincare plot used as confirmative tool in diagnosis of lv diastolic dysfunction for diabetic patients, with and without hypertension. Int. J. Sci. Eng. Res. **4**(10) (2013). ISSN 2229–5518
6. Seyd, P., Ahamed, V., Jacob, J., Joseph, P.: Time and frequency domain analysis of heart rate variability and their correlations in diabetes mellitus, world academy of science, engineering and technology international. J. Med. Health Biomed. Bioeng. Pharm. Eng. **2**(3) (2008)
7. https://www.mathworks.com/matlabcentral/fileexchange/26546-approximate-entropy
8. Fausta, O., Acharya, R., Molinarib, F., Chattopadhyayc, S., Toshiyo, T.: Linear and non-linear analysis of cardiac health in diabetic subjects. Biomed. Signal Process. Control **7**, 295–302 (2012)
9. Corrales, M., Torres, B., Esquivel, A., Salazar, M., Orellana, J.: Normal valueition-sheart rate variability at rest in a young, healthy and active Mexican population. Health **4**, 377–385 (2012)
10. Tarvainen, M., Niskanen, J., Lipponen, J., Ranta-aho, P., Karjalainen, P.: Kubios HRV - heart rate variability analysis software. Comput. Methods Programs Biomed. (2013). https://doi.org/10.1016/j.cmpb.2013.07.024

11. Stein, P., Domitrovich, P., Kleiger, R.: Including patients with diabetes mellitus or coronary artery bypass grafting decreases the association between heart rate variability and mortality after myocardial infarction. Am. Heart J. **147**(2) (2004)

12. Shirole, U., Joshi, M., Desai, K., Bagul, P.: Cardiac autonomous function assessment in congestive heart failure using HRV analysis. Int. J. Scientic Eng. Res. **8**(11) (2017). ISSN 2229-5518

13. Acharya, R., Ghista, D., Subbhuraam, V.: Sudden cardiac death prediction based on nonlinear heart rate variability features and SCD index. Appl. Soft Comput. (2016)

14. Soydana, N., Bretzel, R., Fischer, B., Wagenlehnerb, F., Pilatz, A., Linna, T.: Reduced capacity of heart rate regulation in response to mild hypoglycemia induced by glibenclamide and physical exercise in type 2 diabetes. Metab. Clin. Exp. **62**, 717–724 (2013)

15. Joshi, M., Desai, K., Menon, M.: ECG signal analysis used as confirmative tool in quick diagnosis of Myocardial Infarction. Int. J. Sci. Eng. Res. **3**(3) (2012). ISSN 2229–5518

16. Garciaa, C., Oterob, A., Vilac, X., Marqueza, D.: A new algorithm for wavelet-based heart rate variability analysis. Biomed. Signal Process. Control **8**(6), 542–550 (2013)

17. Joshi, M., Desai, K., Menon, M.: Correlation between Heart Rate Variability and Left Ventricular Ejection Fraction (LVEF) for diabetics and diabetics with hypertension. J. Bioeng. Biomed. Sci. **ISSN**, 2155–9538 (2015)

18. Montano, N.: Heart rate variability explored in the frequency domain: a tool to investigate the link between heart and behaviour. Neurosci. Biobehav. Rev. **33**, 71–80 (2009)

19. Cornforth, D., Jelinek, H.: Detection of congestive heart failure using renyi entropy. Comput. Cardiol. **43**, 667–669 (2016)

A Study on Discontinuity Pattern in Online Social Networks Data Using Regression Discontinuity Design

K. Sailaja Kumar[✉], D. Evangelin Geetha, and T. V. Suresh Kumar

Department of Computer Applications,
M S Ramaiah Institute of Technology, Bangalore, India
sailajakumar.k@gmail.com

Abstract. The analysis of Online Social Networks (OSNs) data is an emerging field involving sociology, statistics, and graph theory. Regression Discontinuity Design (RDD) is a quasi-experimental research design widely used in social, behavioral and related sciences. In this paper, we proposed a methodology to analyze the data from the most popular micro-blogging OSN 'Twitter'. The methodology is implemented using 'R' statistical tool. The tweets related to the 'Mangalayan' event, India's Mars Orbiter Mission launched on 5 November 2013 by the Indian Space Research Organization are analyzed. The Twitter users who are expressive/non expressive on this event are examined. In particular the pattern related to the user's responses to this event is identified, which helps in predicting the Twitter users' social behavior and their involvement associated to such similar events. The most frequent words reflecting the relevance to this event are visualized. The visual results are helpful to understand the pattern or trend of tweets generated by the Twitter users. The users and their tweets in the study are analyzed as two groups based on the word frequency and their relevance to the event. This helps in analyzing the discontinuity pattern in the tweets and exploits the inherent randomness that exists in the frequency of word occurrence using RDD. It is realized from the experimental study that the RDD estimates and plots are credible to analyze the data from the Twitter OSN. Further, it will help the research community to explore the dynamic behavior of the Twitter users adopting this methodology.

Keywords: Indian Space Research Organization (ISRO) · Mangalayan
Online Social Networks (OSN) · Regression Discontinuity Design (RDD)
Twitter · Wordcloud

1 Introduction

The analysis of Online Social Networks (OSNs) data is an inherently interdisciplinary academic field which emerged from social psychology, sociology, statistics, and graph theory. OSNs analysis is now one of the major paradigms in contemporary sociology, and is also employed in a number of other social and formal sciences. Twitter is a public OSN which contains data as Tweets which are used to communicate with other users. These Tweets contain text as well as images or videos, which are also embedded

© Springer Nature Singapore Pte Ltd. 2019
L. Akoglu et al. (Eds.): ICIIT 2018, CCIS 941, pp. 141–150, 2019.
https://doi.org/10.1007/978-981-13-3582-2_11

in Tweets as links. Users have to create small and comprehensive sentences in order to keep up with twitter's length of text restrictions. It is considered as a massive social networking site tuned towards fast communication. Twitter's speed and ease of publication have made it an important communication medium for people from all walks of life. Twitter provides APIs (application programming interface) to access twitter interactions and features. Data pertaining to an event can be obtained from Twitter through these Twitter APIs.

Regression discontinuity design (RDD) is a widely used quasi-experimental research design used in social, behavioral and related sciences. RDD helps to understand the discontinuity or displacement of the regression line at a given point in an assignment variable (cut-off point) that differentiates those in the treatment group from those in the control group. The assignment variable is a continuous variable which is used to designate individuals above or below a set cut-point for the intervention. In this paper we used RDD to study Twitter data related to 'Mangalayan' event.

'Mangalayan' is a space probe, India's Mars Orbiter Mission launched on 5 November 2013 by the Indian Space Research Organization (ISRO) from Sriharikota, Andhra Pradesh. With this India created space history by becoming the first country in the world to enter the Mars orbit in its maiden attempt. This is a huge breakthrough for India in Space Research and this event interested lots of Indians and Non-Indians, to use twitter to convey their emotions online. The Tweets which were relevant to the event were collected and analyzed. In this process, Tweets which are incorrect or posted by the users with less vocabulary in the English language are also observed. Further, these tweets may not be correct either syntactically or semantically. This results in the discontinuity pattern, because of the poor syntax and hence the Twitter API may not be able to identify the tweet as being related to the event and might possibly skip the Tweet. Using RDD technique one can treat these tweets that are not exactly as expected and may involve a huge set of users who have posted these tweets. The tweets in the study are analyzed as two groups based on the cut-off obtained from the word frequency and relevance to the event using 'R' software.

'R' is statistical software, provides the ability to download and analyze Twitter data using the built in R packages. This powerful environment runs on several platforms and is widely used for statistical computing and data analysis. User can extract information on hash tags and can quickly perform visual analysis using the keywords that are found in the tweets. Several R packages are widely used in RDD to identify the treatment effects based on the cut-off and to produce the data-driven RD plots.

In this paper Tweets related to the 'Mangalayan' event are analyzed using visual graphs. The visual results are helpful to understand the pattern or trend of tweets generated by the Twitter users. This also helps to understand the social behavior of the Twitter user and their involvement associated to the event. Further the discontinuity pattern that exists in the words extracted from the tweets related to the 'Mangalayan' event is studied using RDD.

The rest of the paper is organized as follows. Section 2 discusses the related work. Basic concepts pertaining to Twitter and RDD used for Twitter data analysis are presented in Sect. 3. In Sect. 4 methodology to collect, preprocess and analyze the tweets is described. Experimental design, results and discussion are given in Sect. 5. The paper concludes with research future directions in Sect. 6.

2 Related Work

OSN is a social structure involving individuals or organizations as actors and the social interactions (ties) between the actors. The study of social network structures covers the methods for analyzing the network structures as well as variety of patterns observed in these structures. This study also includes social network analysis to identify the influential actors, the local and global patterns and inspect the dynamics of the network [18].

The authors in [15] mentioned, social network analysis is an interdisciplinary academic field arisen from social psychology, sociology, statistics, and graph theory. They also mentioned how Georg Simmel first emphasized the dynamics of social network triads and group associations. The first sociogram was created by Jacob Moreno in the year 1930 to study the interpersonal relationships and in the year 1950 these approaches were mathematically formalized. Later by 1980 the theories and methods of social networks became pervasive in the social and behavioral sciences [8, 18]. However OSN analysis is now one of the major paradigms and is considered to be part of the emerging field of social network analysis [3, 7]. The most popular OSNs are facebook, youtube, Twitter, Linkedin, Google+, Instagram, flicker etc.

Twitter is the most powerful microblogging service which has gained worldwide popularity by more than 500 million users, posting 340 million tweets a day. Scholars used Twitter to predict the occurrence of disasters like earthquakes in the regions like China and Chile. They also identified the relevant users on Twitter to obtain disaster related information [12]. The studies have shown that Twitter data plays a major role on disaster management, mostly on disaster respond than disaster relief or post event [10]. Twitter is used in educational activities to enhance communication building and critical thinking among students. It is also observed that graduates apparently use Twitter to connect with other students and related content which in turn help them to grow professionally and personally [5].

'R' is one of the most versatile open source statistical computing environments used in analyzing the OSN data. Many packages are built in 'R' distribution and many more are available at CRAN website used to extract and analyze the Twitter data (tweets). The packages like *ROAuth, bitops, rjson, Rcurl and streamR* are used to streaming the real-time tweets [1, 4, 6, 9, 13]. It is used to extract and visualize Twitter data using the packages *twitteR, tm, Snowball, RWeka, rJava and RWekajars, ggplot2 and wordcloud*. The packages *rdrobust* and *rdplot*, are used to identify the treatment effects based on the cut-off and to produce data-driven RD plots.

RDD is used to access the effect of a treatment condition by looking for a discontinuity in regression lines between individuals who score lower and higher than some predetermined cutoff score on an assignment variable. The authors [2] used the RDD to determine whether a program for gifted students enhanced their achievement more than regular school placement. The cut-off score to be admitted into the gifted program was set at two or more standard deviations. This is considered to be a strong design, and methodologists have, for a number of years, been trying to get researcher to use this design more frequently.

3 Basic Concepts

3.1 Twitter

Twitter is the most powerful microblogging service created by Jack Dorsey, Evan Williams, Biz Stone, and Noah Glass in 2006 to share the short messages called "tweets" of 140-characters including images, short videos and links. Only registered users can post the tweets, and others can view them. Users can access Twitter through the website interface, SMS or mobile device app. Data pertaining to an event can be obtained from Twitter through Twitter APIs which can be accessed with Twitter user credentials via Open Authentication (OAuth). APIs to access Twitter data can be classified into two types based on their design and access method: [16]

- REST APIs which uses pull strategy for data retrieval. To collect information a user must explicitly request it.
- Streaming APIs uses the push strategy for data retrieval. Once a request for information is made, the Streaming APIs provide a continuous stream of updates with no further input from the user.

The REST API enables developers to access some of the core primitives of Twitter that includes timelines, status updates, and user information. In addition to offering programmatic access to the timeline, status, and user objects, this API also enables developers a multitude of integration opportunities to interact with Twitter. Through the REST API, the user can create and post Tweets back to Twitter, reply to Tweets, favorite certain Tweets, retweet other Tweets, and more [16].

3.2 Regression Discontinuity Design (RDD)

RDD is a widely used quasi-experimental research design in social, behavioral and related sciences. The RDD is named after the discontinuity or displacement of the regression line at a given point in an assignment variable that differentiates those in the treatment group from those in the control group. The assignment variable is a continuous variable which is used to designate individuals above or below a set cut-point for the intervention.

Consider the pre-treatment relationship between the assignment score (S) and the outcome variable (Y) is given by the following straight line linear regression (1):

$$Y = \alpha + \beta S + \varepsilon \tag{1}$$

where α and β are regression coefficients, and ε is the error term.

RDD is an evaluation mechanism used to measure the impact of a treatment assignment mechanism based on the assignment variable. RDD requires an assignment strategy to handle the treatment as well as control groups and this strategy must be known in advance. Objects are assigned to the treatment group and control group base on the cutoff score. If the assignments are proper then the design is called "sharp". On the other hand, if there are some improper assignments or treatment crossovers, the design is called "fuzzy" [11]. An additional advantage of the RDD is its ability to provide unbiased estimates without the need for additional background information [14]. RDD requires information on the assignment variable, the classification into the treatment and control group, and the outcome. This pre-post, two group design is used to assess the outcome of an event or treatment. In this paper, RDD is used to study the Twitter data related to 'Mangalayan' event. The users and their tweets in the study are analyzed as two groups based on the word frequency and relevance to the event.

4 Methodology

Methodology is developed to analyze the data generated by the Twitter related to an event. The steps involved in the analysis are given below

Install R free software using RStudio to collect the Tweets related to the event in the required format from the Twitter:
In order to extract tweets, create a Twitter user account. Use this Twitter user login details to sign in at Twitter Developers and create a Twitter application.

Install and load the necessary R packages:
ROAuth: It provides an interface to the OAuth 1.0 specification, allowing users to authenticate via OAuth to the server of their choice. The authentication of API requests on Twitter is carried out using OAuth.

twitteR: It provides an interface to the Twitter web API. A user's Tweets can be retrieved using both the REST and the Streaming API.

Extract Tweets related to the event for a specific period using the search string:
Use the function *searchTwitter()* to search Twitter based on the search string and then obtain the tweets in text format. Twitter provides the search/tweets API to facilitate searching the Tweets. The search API takes words as queries and multiple queries can be combined as a comma separated list. Tweets from the previous ten days can be searched using the REST API.

Analyze the Tweet count:
Obtain the tweets count day wise to observe the distribution of tweets over a period related to the event. Generate a line graph using the *plot()* function in R which is used to visualize the results to understand the pattern or trend of tweets generated by the Twitter users. This also helps to understand the social behavior of the Twitter user and their involvement associated to the event.

Preprocess the Tweets for analysis:
Convert the tweets into a data frame and then to a corpus, which is a collection of text documents. The corpus is then processed with the functions provided in R package *tm*. The corpus is then transformed to new corpus by changing letters to lower case, and removing punctuations, numbers, hyperlinks, stop words and replace special characters "/", "@" and "|" with space. Transformation is performed using *tm_map()* function.

Another important preprocessing step is to reduce words to their root form. In other words, this process removes suffixes from words to make it simple and to get the common origin. Stem the extracted words to retrieve their radicals, so that various forms derived from a stem would be taken as the same when counting word frequency. Word stemming can be done using the R packages like; *Snowball, RWeka, rJava and RWekajars*. Further complete the stems to their original forms using the function *stemCompletion()*.

Build a term-document matrix from the corpus obtained after stemming using the function *TermDocumentMatrix()*. Term-document matrix represents the relationship between terms and documents, where each row stands for a term (word) and each column for a document (tweet), and an entry is the number of occurrences of the term in the document.

Analyze the Tweets:
Find frequent words from term-document matrix: Frequent words are found from the term-document matrix using the function *findFreqTerms()*. From the term-document matrix, the frequency of terms (words) can be derived using the function *rowSums()*. Select the word that appears in two or more tweets.

A barplot is constructed using the R package *ggplot2* to visually represent the frequency of words. Wordcloud is generated using the R package *wordcloud*, which is used to quickly visualize the important words extracted from the tweets related to the event. It also adds simplicity and clarity. A word cloud is a text mining method that allows us to highlight the most frequently used keywords in a paragraph of texts. It is also referred to as a text cloud or tag cloud.

Identify the discontinuity pattern in the data:
Identify the discontinuity pattern in the words and also in the tweets related to the event using RDD. The functions *rdrobust()* and *rdplot()* available in R package *rdrobust* [17] which provides a data driven graphical inference procedures for RDD.

5   Experimental Design, Results and Discussion

The experimental setup to implement the methodology described in Sect. 4 is given below. The experimental is used to analyze the discontinuity in the tweets generated related to an event.

System Configuration

- Intel® Pentium® D CPU 3.40 GHZ
- RAM: 2.99 GB, 32 bit operating system
- Internet Speed: 100 Mbps
- Software tools: R packages and Twitter APIs

  Data collection duration: 23/9/2014 to 29/09/2014.

5.1   Collect Required Data from Twitter Using Suitable API

Tweets were extracted related to 'Mangalayan' event during the period 23/09/2014 to 29/09/2014 using the Twitter REST API by constructing the necessary queries. This Twitter API facilitates the search by taking the event related words as a query. The extracted tweets are stored in a .csv file.

5.2   Analyze the Tweet Count

The day wise distribution of tweets (tweet count) during the period 23/09/2014 to 29/09/2014 related to 'Mangalayan' event is obtained and is listed in Table 1. From Table 1 it is observed that number of tweets related to the event is high during the dates nearer to the event happened and gradually decreases in later dates. This is in general the social behavior of the Twitter users in response to an event. It is also observed that 75% users shown interest within one week and the trend was reduced after 8 days of the event. This observation is clearly visualized in the line graph presented in Fig. 1. The line graph helps to determine the important time periods, where the users' response to the event is high and also to identify the trend in tweet generation.

Table 1. Tweets count for the period 23/09/2014 to 29/09/2014

23/9/2014	24/9/2014	25/9/2014	26/9/2014	27/9/2014	28/9/2014	29/9/2014
245	293	218	135	103	13	15

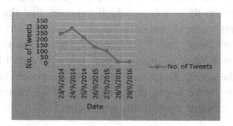

Fig. 1. Distribution of tweets day wise

5.3   Preprocess the Tweets

The TermDocumentMatrix is constructed and the output is presented in Fig. 2. From Fig. 2 it is observed that there are 1839 terms and 1027 documents (tweets). The 99% sparsity means 99% of the entries are 0, so the row does not exist in a particular document column. The term occurring in all the tweet documents is 'isro' and the terms with more than 100 occurrences are shown in Fig. 3. The observation from Fig. 3 is, some of the top terms relevant to the event are 'mangalyan', 'mangalayan', 'mars', 'scientist'.

```
A term-document matrix (1839 terms, 1027 documents)

Non-/sparse entries: 11902/1876751
Sparsity            : 99%
Maximal term length: 75
weighting           : term frequency (tf)
```

Fig. 2. Term-document matrix

achiev	first	great	india	isro	mangalayan	mangalyaan	
30	525	599	707	750	909	919	
mar	mission	news	orbit	proud	reach	scientist	success
933	1001	1074	1110	1210	1248	1351	1499

Fig. 3. Distribution of terms with > 100 occurrences

5.4 Analyze the Tweets

The distribution of the word frequencies that is the association between the word and the word frequency is visualized using the Bar plot given in Fig. 4. Here the most frequent words related to the event "Mangalayan' are ordered alphabetically. Figure 4 explains the distribution of word frequencies which confirms to the proper character-istic of the collected corpus. The sample of the most frequent words are: 'mangalayan, 'mangalyaan', 'isro', 'india', 'mars' etc., The most frequent words reflecting the rel-evance to the event is visualized using a word cloud as shown in Fig. 4. In the word cloud, the size of the keyword represents how often the word occurs in the tweet documents. The word cloud will only display the most popular words in the tweets and don't show how the words are related. From Fig. 4 it is observed that 'mangalayan', 'isro', 'india', are the top three words, which validates that the search string 'man-galayan' based tweets present information on 'mangalayan' event. Some other important words are 'scientist', 'news', 'success', 'mar', 'mission', which shows that it focuses on documents related to 'mangalayan' event. Words like 'modi' and 'rad-hakrishnan' reflects the tweets about congratulating the people behind the success of the event. There are also some tweets on the expenses of the 'mangalayan', as indicated by words 'cheapest', 'too low cost' in the world cloud.

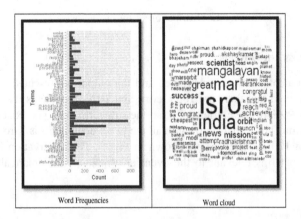

Word Frequencies   Word cloud

Fig. 4. Word frequencies and word cloud of 'Mangalyaan' event

5.5 Identify the Discontinuity Pattern in the Twitter Data

Identifying the discontinuity pattern that exists in the user's response to the event 'Mangalayan, helps in predicting the Twitter users' social behavior. To identify the discontinuity pattern that exists in the tweets obtained related to the event 'Mangalayan', the variables Word-Frequency-Count (WQC) and Word-Significance (WS) are considered. The variable WFC records the occurrence of a particular word, while the variable WS records the variation of the cut-off value from the occurrence of a particular word frequency. This cut-off is used as an assignment criterion to assign the words to the treatment. This assignment is based on random factor which is unmeasurable or unobservable. The different cut-off values considered for the analysis are the words frequency 50, 100, 150, 200 and 250. An X-Y plot is generated using the function supported by R which shows the relationship between the two variables WFQC and WS and the results are presented in Fig. 5. From the regression graphs it can be observed that there exists discontinuity in the word relevance pattern. Moreover, the discontinuity pattern is highly visible as the cut-off value is increasing. The words can be assigned for the treatment based on their frequencies above or below the known cutoff. That is the words which are having high frequency are assigned to the control group and words having low frequency are assigned to the treatment group. From the regression graphs it is observed that the words which are below the cut-off are having less frequency but they are related to the event 'Mangalayan'. In addition, the RDD also compares the users' behavior with respect to showing and not showing interest on the event 'Mangalayan'. This is used to identify the casual behavior of the user. It is based on the assumption that the words closed to the regression margin represent the users showing interest in the event and is determined by the random factor which is unmeasurable or unobservable.

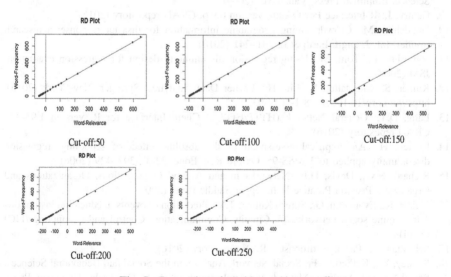

Fig. 5. Regression plot with varying cut-off

6 Conclusion and Future Work

In this paper, Twitter users who are expressive/non expressive on the event 'Mangalayan' are examined. In particular the pattern related to the user's response to the event are identified which helps in predicting the Twitter users' social behavior. The most frequent words reflecting the relevance to the event are visualized using a wordcloud. Further, it helps to identify the discontinuity pattern exists in the words using RDD. RDD exploits the inherent randomness that exists in the frequency of word occurrence. It is realized from the experimental study that the RDD estimates are credible to analyze the data related to an event obtained from the Twitter.

References

1. Barbera, P.: Access to Twitter Streaming API via R, version 0.2.1, CRAN repository (2014)
2. Braden, J.P., Bryant, T.J.: Regression discontinuity designs: applications for school psychology. School Psychol. Rev. **19**(2), 232–239 (1990)
3. Borgatti, S.P., Mehra, A., Brass, D.J., Labianca, G.: Network analysis in the social sciences. Science **323**(5916), 892–895 (2009)
4. Couture-Beil, A.: Package 'rJson', version 0.2.15, CRAN repository (2015)
5. Domizi, D.: Microblogging to foster connections and community in a weekly graduate seminar course. J. TechTrends **57**(1), 43–51 (2013)
6. Dutky, S., Maechler, M., Dutky, S.: Package 'bitops', version 1.0-6, CRAN repository (2016)
7. Easley, D., Kleinberg, J.: Overview, Networks, Crowds, and Markets: Reasoning About a Highly Connected World, pp. 1–20. Cambridge University Press, New York (2010)
8. Freeman, L.: The Development of Social Network Analysis: A Study in the Sociology of Science. Empirical Press, Vancouver (2004)
9. Gentry, J.: R Interface For OAuth, version 0.9.6. CRAN repository (2015)
10. Goodchild, M.: Crowdsourcing geographic information for disaster response: a research frontier. Int. J. Digit. Earth **3**(3), 231–241 (2010)
11. Hyunshik, L., Tom, M.: Using regression discontinuity design for regression evaluation. JSM (2008)
12. Kumar, S., Morstatter, F., Liu, H.: Twitter Data Analytics. Springer, New York (2013). https://doi.org/10.1007/978-1-4614-9372-3
13. Lang, D.T.: General Network (HTTP/FTP/...) Client Interface for R. version 1.95-4.8. CRAN repository (2016)
14. Luyten, H.: An empirical assessment of the absolute effect of schooling: regression discontinuity applied to TIMSS-95. Oxford Rev. Educ. **32**(3), 397–429 (2006)
15. Richard, S.W., Davis, D.F.: Networks In and Around Organizations. Organizations and Organizing. Pearson Prentice Hall, Upper Saddle River (2003)
16. Sailaja, K., Evangelin, G., Suresh Kumar, T.V.: Prediction of events in education institutions using online social networks. In: Circuits, Communication, Control and Computing (I4C) (2014)
17. Sebastian, C.: Package rdrobust. CRAN repository (2016)
18. Stanley, W., Katherine, F.: Social Network Analysis in the Social and Behavioral Sciences. Social Network Analysis: Methods and Applications, pp. 1–27. Cambridge University Press, Cambridge (1994)

"Senator, We Sell Ads": Analysis of the 2016 Russian Facebook Ads Campaign

Ritam Dutt[1], Ashok Deb[2(✉)], and Emilio Ferrara[2]

[1] Indian Institute of Technology Kharagpur, Kharagpur 721302, India
[2] University of Southern California, Los Angeles, CA 90089, USA
adeb@usc.edu

Abstract. One of the key aspects of the United States democracy is free and fair elections that allow for a peaceful transfer of power from one President to the next. The 2016 US presidential election stands out due to suspected foreign influence before, during, and after the election. A significant portion of that suspected influence was carried out via social media. In this paper, we look specifically at 3,500 Facebook ads allegedly purchased by the Russian government. These ads were released on May 10, 2018 by the US Congress House Intelligence Committee. We analyzed the ads using natural language processing techniques to determine textual and semantic features associated with the most effective ones. We clustered the ads over time into the various campaigns and the labeled parties associated with them. We also studied the effectiveness of Ads on an individual, campaign and party basis. The most effective ads tend to have less positive sentiment, focus on past events and are more specific and personalized in nature. The more effective campaigns also show such similar characteristics. The campaigns' duration and promotion of the Ads suggest a desire to sow division rather than sway the election.

Keywords: Social media · Information campaigns · Ads manipulation

1 Introduction

One of the key aspects of the United States democracy is free and fair elections, unhindered by foreign influence, that allow for a peaceful transfer of power from one President to the next. Campaign Finance laws forbid foreign governments or individuals from participating or influencing the election. The 2016 US presidential election stands out not only due to its political outsider winner, Donald J. Trump, but also due to suspected foreign influence before and during the election. It is alleged that the Russian Federation operated the Main Intelligence Directorate of the General Staff, a military intelligence agency. This agency is suspected of influencing the election with resources allocated towards social media on a variety of forums.

© Springer Nature Singapore Pte Ltd. 2019
L. Akoglu et al. (Eds.): ICIIT 2018, CCIS 941, pp. 151–168, 2019.
https://doi.org/10.1007/978-981-13-3582-2_12

Corporate leadership and council from Google, Twitter, and Facebook testified on November 1, 2017 to the Senate Intelligence Committee concerning *social media influence* on their platforms. While Facebook's General Council suggested that it would be difficult to verify that every ad purchased on their platform adheres to US campaign finance laws, intuitively ads purchased in Russian Rubles would be highly suspicious. There were approximately 3,500 ads identified by Facebook that met such criteria totaling close to $100K and purchased between June 2015 and August 2017. The surfacing of these ads contributed to Facebook's CEO Mark Zuckerberg's testimony of 10–11 April 2018 to a number of Senate and House Committees.[1]

Rep. Adam Schiff, the ranking member of the House Intelligence Committee, had voiced his opinion that the Russians had launched an 'independent expenditure campaign on Trump's behalf, regardless of his involvement.[2] However, as emphatically stated by Rob Goldman,[3] the vice president of ads at Facebook, the over-arching aim of the advertisements was to bring about discord among different communities in the United States. In Goldman's words, "(the ads) sought to sow division and discord" in the political proceedings before, during, and after the 2016 US elections by leveraging the freedom of free speech and pervasive nature of social media. This statement is contradictory to the claim that the primary objective of the advertisements was to influence the effect of the 2016 elections and sway it in favor of Trump or to vilify Clinton. Under the direction of Democrats on the House Intelligence Committee, the Russian Facebook ads were released to the public on May 10, 2018. The main objective of this work is to apply language analysis and machine learning techniques to understand the intent of those ads by exploring their effectiveness from a campaign perspective.

2   Related Literature

Since the early 2000s, there has been increasing research in the new domain of *computational social science* [21]. Most of the literature has focused on networked influence, information (or misinformation) diffusion, and social media association with real-world events [16]. As it concerns our research efforts, related work focuses on social media use in politics as well as campaign detection.

Politics in Social Media: Concerning divisive issue campaigns on Facebook, ongoing work has explored the organizations and objectives behind the Russian ads from a political communication standpoint. Kim [19] stated that suspicious groups which could include foreign entities are behind many of the divisive campaigns. Additionally, approximately 18% of the suspicious groups

---

[1] https://www.judiciary.senate.gov/meetings/facebook-social-media-privacy-and-the-use-and-abuse-of-data.

[2] http://www.businessinsider.com/russian-facebook-ads-2016-election-trump-clinton-bernie-2017-11.

[3] https://www.cnbc.com/2018/02/17/facebooks-vp-of-ads-says-russian-meddling-aimed-to-divide-us.html.

were Russian. The authors asserts that there are shortcomings in federal regulations and aspects of digital media that allow for anonymous groups to sow division [19]. While Kim approaches the issue from a policy perspective, we focus more on the effectiveness and organization of the ads themselves. While the data we used in this research is specific to only the Russian Facebook ads, we present a methodology that could be extended to automatically sort any ads into their divisive campaigns. Previous work established that social media platforms were exploited during the 2016 US Presidential Election [1, 2, 4, 39], as well as numerous other elections [9, 12, 17, 24, 28, 34] and other real-world events [11, 13], by using tools like bots and trolls [8, 14, 37].

Campaign Detection: In order to combat misinformation, it is necessary to understand the characteristics that allow it to be effective [23]. In addition to misinformation, divisive information which creates polarized groups is counter to what the political system or a democratic nation needs to thrive [35]. Previous campaign detection has been focused on spam [10] and malware [20, 33] in order to protect computer information systems. The most relevant work for campaign detection on social media is by Varol and collaborators [15, 38]. They use supervised learning to categorize Twitter memes from millions of tweets across a series of hashtags. In comparison to that work, we focus at the microscopic level on paid Facebook ads determined to be from the same source. In addition to looking at the Russian campaign messaging and content, we are able to factor cost and effectiveness into our analysis.

3   The Data

3.1   Collection

The dataset comprises 3,516 advertisements with 22 variables as released by the Data for Democracy organization in csv format.[4] The data was released under the direction of the Democrats on the House Intelligence Committee. The ads were released to the public on May 10, 2018. The ads were purchased in Russian Rubles during the 2016 US Presidential election and beyond from June 2015 to August 2017. In analyzing effectiveness, we only considered ads which were viewed by at least one person (impressions greater than zero). In analyzing campaigns, all ads with non-zero impressions or those which were purchased in Rubles were considered. Our dataset consists of the Russian Facebook paid ads totaling $93K. Again, the data was initially provided by Facebook, so there is no way to independently verify its completeness and ads purchased in Rubles would be a lower bound to all ads purchased on behalf of the broader operation. Summary statistics of the data are shown in Table 1.

3.2   Preliminary Data Analysis

Clicks and Impression Counts: Clicks and impressions are important metrics to understand the outreach and efficacy of the advertisement. Clicks, or

---

[4] https://data.world/scottcame/us-house-psci-social-media-ads.

Table 1. Summary statistics of Russian Facebook ad dataset

Criteria	Count	Value	Analysis
All ads	3,516	$100K	Individual
Ads with at least 1 impression	2,600	$93.0K	Effectiveness
Ads with at least 1 impression AND paid in RUB	2,539	$92.8K	Campaign & Party

link clicks, quantify the number of people who have clicked on the ad and was redirected to the particular landing-page. Impressions refer to the total number of times the advertisement has been shown. It differs from Reach which reflects instead the number of individual people who have seen the ad. We present the distribution of impressions and clicks for the FB ads in Figs. 1a and b, respectively. It is clearly evident that a majority of ads have attained sufficient outreach and popularity. We observe that a huge fraction of the ads are targeted to the younger age group as seen in Fig. 1c.

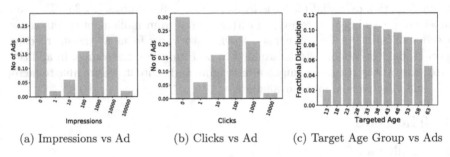

(a) Impressions vs Ad   (b) Clicks vs Ad   (c) Target Age Group vs Ads

Fig. 1. Distribution by impressions, clicks and target age group

4   Research Framework and Research Questions

In this paper, we define our research framework tackling the following problem:

Research Question: What features are associated with the engagement of the Russian Facebook ads and what was their impact (i.e., *how effective were they*) at a campaign-wise (*operational*), and on a party-wise (*strategic*) basis?
To operationalize this question we split it into three sub-parts:

1. What features are associated with the engagement of the Russian Facebook ads.
Definition of engagement: To quantify engagement, we estimate how likely a person would respond to an ad when it is shown to them. The metric we use is Click-Through Rate (CTR). We approach the problem by classifying ads which have non-zero impressions in two groups, namely *more effective* and *less effective* ads. The classification is done using a decision rule where the median value of the CTR across all ads is the threshold. We consider non-zero impressions only since we cannot evaluate the effectiveness of an ad that was not seen.

We then analyze the stylistic and textual features of the Ads between the two categories, using different natural language processing techniques. The features include sentiment, emotion, structural content, parts of speech distribution, named entity distribution, and linguistic categories. We note those features which show significant differences across categories.

2. What was the Ads impact (effectiveness) at a campaign-wise (operational) level? **Definition of effectiveness**: At the campaign level we define effectiveness as the audience reach efficacy. The metric we use is Cost Per Thousand Impressions (CPM) and Cost Per Click (CPC) (explained in methodology). We approach this question by clustering the ads into the various campaigns and using CPM to determine the most and least effective campaigns as well as any insights from the associated features mentioned in the first sub-part.

3. What was the Ads impact (effectiveness) at a party-wise (strategic) level? **Definition of effectiveness**: We define the effectiveness at the party level by observing significant differences in terms of CTR, CPM and total cost between the parties. We create parties by manually labeling the ads into Democratic (Blue), Republican (Red) and Neutral (Green). We exclude the Neutral campaigns and assess the effectiveness of the Blue and Red parties and report any significant findings from a feature-wise perspective.

It is notable to mention our assumption that all of these ads within the campaigns and parties were generated by the same alleged organization in Russia.

5   Methodology

5.1   Features of Effective Ads

The effectiveness of ads at the individual level is measured using CTR.

Click-Through Rate (CTR). CTR of a particular advertisement is the ratio of clicks to impression for the ad expressed as a percentage. CTR reflects the creativity and compelling nature of the advertisement [7].

$$CTR = \frac{\#Clicks}{\#Impressions} * 100(\%) \qquad (1)$$

The stylistic and textual features associated with the ads we analyzed are:

Sentiment Analysis: Sentiment analysis helps to identify the attitude of the text and gauge whether it is more positive, negative or neutral. Based on the comparative analysis of in [30], we utilized 2 methods to determine sentiment on the Ad text to obtain the overall compounded sentiment score of the Ad. VADER: Valence Aware Dictionary for Sentiment Reasoning [18] is a rule-based sentiment model that has both a dictionary and associated intensity measures. Its dictionary has been tuned for micro-blog like contexts. We also observe the categories corresponding to positive and negative emotions by performing LIWC [27] analysis on the *Ad Text*.

Emotion Analysis: We leverage the NRC lexicon of [26] to calculate the average number of words corresponding to an emotion per advertisement. The associated 8 emotions include anger, anticipation, joy, fear, trust, disgust, sadness and surprise Fig. 2.

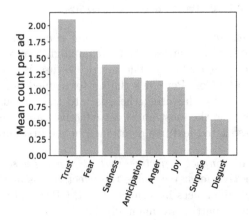

Fig. 2. Emotion word counts

Structural Content: The structural content of the text refers to the *distribution of sentences and words* per advertisement. An ad's efficiency often correlates with the amount of textual content [31].

POS-TAG Distribution: We employed the inbuilt Part-of-Speech (POS) tagger of NLTK [36] and the Penn Tree Bank[5] to observe the distribution of different POS (Parts of Speech) TAGS in the advertisement texts.

Named Entity Recognition (NER) Distribution: The high proportion of proper nouns from POS TAG analysis signifies that the ads cater more to real-world events. Consequently, we also inspected the distribution of different named entities using the Perceptron-based NER of Stanford CoreNLP [22] pertaining to "PERSON", "ORGANIZATION" and "LOCATION".

Linguistic Inquiry and Word Count (LIWC) Analysis: LIWC [27] computes the proportion of words in psychologically meaningful categories for the analyzed text which we leverage to discover different linguistic and cognitive dimensions.

5.2   Campaign-Level Analysis

We leverage different methods to cluster Ads into non-overlapping campaigns:

LDA Topic Extraction: We implemented LDA [5] (Latent Dirichlet Allocation) using the in-built gensim model of [29] on the corpus of advertisement

---

[5] https://www.ling.upenn.edu/courses/Fall_2003/ling001/penn_treebank_pos.html.

text to obtain a list of 50 topics. However these topics had several overlapping words and dealt with racism, gun-control or police accountability. It also failed to capture broad topics like homosexuality or immigration.

Key-Word/Key-Phrase Extraction: We also employ RAKE (Rapid Automatic Keyword Extraction) [32], an unsupervised, domain-independent and language-independent technique to extract keywords from the advertisement texts. This methodology captures niche topics since it observes each document individually.

However, the above methods suffer two shortcomings. Firstly, they did not take into account the Ad's images which serves the purpose of propagating ideas mentioned in the advertisement. Secondly, these techniques do not incorporate the context associated outside of the text. For example, the ad text, "The blue gang is free to do whatever they want" clearly refers to police brutality, often misdirected at African-Americans, but it is impossible to decipher from the text alone. Hence we resort to a semi-automated network clustering technique to identify campaigns as described below.

Network-Based Clustering: Some of the advertisements have a meta-data field labeled Interests which corresponds to topics. We represent each topic as a vector, obtained by the FastText technique of [25]. We compute similarity between the topics using the given equation:

$$sim(T_i, T_j) = \alpha(\boldsymbol{T}_i \cdot \boldsymbol{T}_j) + (1 - \alpha)\frac{|Ads(T_i) \cap Ads(T_j)|}{min(|Ads(T_i)|, |Ads(T_j)|)} \qquad (2)$$

wherein, T_i and T_j represent two arbitrary topics, $\boldsymbol{T}_i$ and $\boldsymbol{T}_j$ denote their vector representation and $Ads(T_i)$ enlists the ads which have T_i as a topic. The first part of the equation ($\boldsymbol{T}_i \cdot \boldsymbol{T}_j$) simply computes the cosine similarity score of the topic vectors, while the second part calculates the overlap coefficient similarity between two topics. While $\alpha \in [0,1]$ determines the trade-off between the two similarity scores.

Each topic is then represented as a node in an undirected graph with edges representing the similarity between two topics. We binarize the graph by retaining only the edges above a certain threshold β and cluster it. We experimented with different values of α, β and different algorithms and experimentally verified that the Louvain algorithm [6] with a threshold of 0.9 for both α and β gave the best results with 9 non-overlapping campaigns. A change in α and β values drastically altered the number of communities, ranging from 2–3 on one extreme to 40–50 in another. Likewise, ML based unsupervised clustering techniques like KMeans or Spectral Clustering were unable to incorporate the overlap coefficient similarity and hence showed poorer performance.

Thus each topic belongs to one of the initial 9 campaigns. Since an ad can contain several topics, they can belong to different campaigns, we assign them to the campaign that with the most number of topics, breaking ties arbitrarily.

We then manually inspected the rest of the ads and assigned those which did not have the Interests field to one of the 9 campaigns. Sometimes, we had to create new campaigns since the particular Ad did not conform with any of the previous ones. It was necessary to break up large clusters which had similar notions (police brutality, racism and Black Lives Matter) into different campaigns. Eventually, that yielded the final 21 campaigns as demonstrated in Table 2.

The effectiveness of ads at the campaign level is measured via CPM and CPC.

Cost Per Thousand Impressions (CPM): CPM for an ad is simply the amount of Rubles spent to reach a mile (thousand impressions). CPM is primarily determined by the target audience [3].

$$CPM = \frac{AdCost(RUB)}{\#Impressions} * 1000 \tag{3}$$

Cost Per Click (CPC): CPC for an Ad is the amount of Rubles required to receive a click. CPC reflects the traffic generated by the ad to the landing page [3].

$$CPC = \frac{AdCost(RUB)}{\#Clicks} \tag{4}$$

A campaign's effectiveness is usually measured by a *low CPC* value because it implies that the amount of Rubles required to get an audience's response is also less. However a low CPM is sometimes essential if one wishes to target a particular audience and optimize the overall cost of the campaign. If an ad itself has a high CTR, purchasing Ads using CPM may be a better strategy.

The stylistic features analyzed are consistent with those outlined in Sect. 5.1.

5.3 Party Clustering

The campaigns are manually assigned to parties as stated in Table 2.

The effectiveness of ads at the party level is measured using CPC and CPM, in a fashion similar to the campaign-wise analysis.

6 Results

6.1 Ad Effectiveness in Aggregate

The calculated median CTR value of the advertisements is 10.24. Consequently, we categorize the ads as more or less effective if the CTR value is greater or lesser than 10.243 respectively. We present the significant semantic and textual features here. In all cases, significant difference refers to a *p-value* of ≤ 0.001.

Sentiment Analysis: We observe that the overall compounded score is significantly lower for the more effective ads than those in the less effective ads, implying that the former ads tend to be less positive. Surprisingly, there is no significant difference between the distribution of negative sentiments.

Table 2. Campaigns identified in the dataset and parties associated with them.

Campaign	Definition	Party
Police Brutality	Injustice meted out to the Blacks by the Police	Democrat
Entertainment	Multi-media sources of entertainment (memes, songs, videos)	None
Prison	Prison reforms against mandatory sentences, prison privatization	Democrat
Racism	Acts of racism harbored against any racial minority in America	Democrat
LGBT	Rights and dignities for the LGBT people	Democrat
Black Lives Matter	Incarceration, shooting or other acts of cruelty against Blacks	Democrat
Conservative	Ideals of patriotism, preserving heritage and Republican advocacy	Republican
Anti-immigration	Preventing illegal immigration across the US borders	Republican
Veterans	Support for the hapless/ crippled veterans of war	Republican
2nd Amendment	Supporting the right to bear arms and guns	Republican
All Lives Matter	Counter to the Black Lives Matter	Republican
Anti-war	Opposition of wars and acts of aggression against the Middle East	Democrat
Texas	A medley of Conservative Ads specifically leaned towards Texas.	Republican
Islam	Against Islamophobia and support for the Muslims in the US	Democrat
Immigration	Support the immigration of other nationalities into America	Democrat
Liberalism	In support of the various liberal reforms by the Blue part	Democrat
Religious	Support for the conservative Christians in the US	Republican
Hispanic	Support for the Hispanic/ Latino community in the US	Democrat
Anti-Islam	Messaging against the acceptance of Muslims in US	Republican
Native	Support for the Native American Indians and their community	Democrat
Self-Defense	Focused on martial arts training for anti-police brutality	Democrat

Emotion Analysis: None of the 8 emotions showed any significance difference across the two categories, except surprise which demonstrated mild significance ($p\text{-}value \leq 0.05$).

Structural Content: The distribution of sentences and words per advertisement do not vary significantly across the two categories.

POS-TAG Distribution: We observe that adverbs (RB) and past tense verbs (VBD) occur more frequently in the more effective ads. This implies that more effective ads tend to refer to past events more frequently while the pronounced usage of adverbs implies that actions are explained in detail. However, the proportion of nouns across advertisements is very high, with NN (common nouns, singular) and NNP (proper nouns, singular) accounting for 6.32 and 5.89 words per advertisement respectively.

NER Distribution: The NER analysis revealed that the category "PERSON" occurred in significantly higher proportion among the more effective ads.

LIWC Analysis: Only the most significant LIWC categories have been taken to account here.

Personalization: Categories belonging to *SheHe* and *Ipron (personal pronouns)* are higher in more effective ads, while those belonging to *We* and *Friends are lower* in the more effective category. This indicates the more effective ads are more personalized or cater to the individuals rather than the communities.

Table 3. Average values between more effective and less effective. Significance of the feature as denoted by *, **, ***, **** correspond to *p-values* less than 0.05, 0.1, 0.001 and 0.0001 respectively.

Category	Less effective	More effective	Mean diff	T-value
Compounded sentiment***	0.139	0.048	−65.699	3.687
Positive sentiment****	0.19	0.158	−16.61	4.414
Negative sentiment	0.097	0.089	−8.243	−6.281
Surprise**	0.503	0.64	27.4	−2.836
Anger	1.061	1.174	10.648	−1.37
#Sentences	3.768	3.733	−0.939	0.232
#Words	48.008	52.648	9.666	−1.873
RB (Adverb)****	1.454	2.016	38.698	−4.85
VBD (Verb, past)****	0.896	1.414	57.834	−4.927
NN (Common nouns)	6.296	6.555	4.111	−0.776
NNP (Proper nouns)	6.361	6.044	−4.983	0.93
PERSON***	0.017	0.028	61.658	−3.209
LOCATION*	0.012	0.009	−21.195	2.127
Ipron****	0.028	0.04	43.807	−5.973
We****	0.033	0.017	−48.568	8.157
SheHe****	0.005	0.013	148.432	−6.904
Friends****	0.004	0.001	−64.348	4.588
Money****	0.01	0.005	−48.167	5.423
Religion****	0.01	0.004	−61.087	4.424

Religion and Money: *Religion and Money* occur in lower proportions in the more effective ads than the less effective ones. This shows that religious or financial divide are not as successful to ensure engagement.

We now present the differences in Table 3. The columns corresponding to Less Effective(Mean) and More Effective(Mean) specify the mean value of the distribution for the categories. The Mean Diff column is simply computed

$$MeanDiff = \frac{High(Mean) - Low(Mean)}{Low(Mean)} * 100\% \tag{5}$$

The stars beside a category name correspond to the level of significance as indicated by the *p-value*.

6.2   Campaign-Wise Analysis

We present the statistics of the different campaigns in Table 4 which are arranged in decreasing order of their effectiveness and thus in increasing order of CPM. We demarcate the campaigns into more and less effective based on the median value of the CPM (A more effective campaign has a CPM score less than 277.57).

Table 4. Statistics of the campaign arranged in decreased order of effectiveness.

Topics	Cost in RUB	Cost in USD	Frequency	Impressions	Clicks	CPM	CPC	CTR
Hispanic	164,146.40	2,628.05	186	5,943,904	713,804	27.62	0.23	12.01
Immigration	2,971.30	47.76	10	74,344	10,762	39.97	0.28	14.48
All Lives Matter	150,372.36	2,368.50	11	1,890,020	82,779	79.56	1.82	4.38
Black Lives Matter	1,807,407.97	28,631.85	1206	19,273,576	1,856,476	93.78	0.97	9.63
Entertainment	90,188.75	1,407.42	159	885,273	87,956	101.88	1.03	9.94
Racism	237,900.47	3,677.33	125	1,364,627	82,168	174.33	2.9	6.02
Native	9,397.14	160.94	12	47,428	5,355	198.13	1.75	11.29
Religious	212,647.46	3,543.32	21	1,032,898	78,669	205.87	2.7	7.62
2nd Amendment	234,324.96	3,833.16	50	1,119,281	87,986	209.35	2.66	7.86
Police Brutality	563,945.02	8,873.97	194	2,535,621	207,233	222.41	2.72	8.17
Veteran	220,615.91	3,468.31	97	794,826	59,925	277.57	3.68	7.54
Conservative	831,223.67	13,600.98	116	2,773,169	213,894	299.74	3.89	7.71
Anti-Islam	4,385.58	69.64	3	13,949	2,725	314.4	1.61	19.54
LGBT	303,738.01	4,796.96	95	887,058	82,217	342.41	3.69	9.27
Anti-war	27,469.85	444.45	15	75,517	6,980	363.76	3.94	9.24
Islam	271,567.36	4,271.96	56	581,392	22,033	467.1	12.33	3.79
Liberalism	87,405.43	1,387.71	33	177,089	15,542	493.57	5.62	8.78
Texas	295,043.68	4,698.09	35	589,409	51,400	500.58	5.74	8.72
Prison	13,552.58	215.30	19	25,954	1,981	522.18	6.84	7.63
Self-defense	30,982.02	518.22	25	53,712	2,136	576.82	14.5	3.98
Anti-Immigration	289,898.95	4,432.61	71	419,380	57,865	691.26	5.01	13.8

We note the following stylistic differences between the more effective and less effective campaigns.

Sentiment Analysis: The compounded sentiment score is significantly lower for the more effective campaigns since those campaigns involve serious topics like police brutality, racism, etc.

Emotion Analysis: All 8 emotions, barring surprise, are observed to be significantly pronounced in the more effective campaigns. We hypothesize that ads evoking emotions are likely to be shared more and hence the impressions increase for the ad, thereby decreasing the potential CPM.

Structural Analysis: Surprisingly, we note that ads in the more effective campaigns tend to be of shorter length, i.e more concise.

POS-TAG Distribution: POS corresponding to adverbs (RB), plural nouns (NNS, NNPS) and verbs (VB) occur more frequently in the less effective campaigns, the significance of which is unknown.

Named Entity Distribution: Named entity mentions corresponding to 'PERSON' is significantly higher in the more effective campaigns while 'LOCATION' is higher in the less effective ones. This finding is attributed to disproportionate large mentions of victims of racial prejudice in the more effective campaigns. Likewise, the less effective campaigns include Texas, Anti-Immigration to US, Veterans, etc which directly reference America.

LIWC Analysis: In the category of *Religion*, the less effective campaigns have a higher proportion of ads associated with Islam. This conforms the analysis at the individual ad level that religious ads are less effective. As for *Associativity*, the more effective campaigns are also individualistic/personal as opposed to community-driven. This finding is substantiated by the significantly high frequency of *I* and *We* categories respectively in the more and less effective campaigns.

We also observe that the individual CPM and cost spent on an ad is significantly lower for the more effective campaigns than the less effective ones. Likewise, the number of clicks and CTR of an individual Ad is significantly higher for the more effective campaigns. Thus, more effective ads do contribute to effective campaigns, although the effectiveness metrics themselves are different for the parties and campaigns.

6.3 Party-Wise Analysis

We present a statistical overview of the ads of the two parties in Table 5.

Table 5. Performance of the two parties.

Party	# Ads	Cost	Cost	Clicks	Impressions	CPM (RUBs)	CPC (RUBs)	CTR
Democrat	1,976	3.5MRUB	$55.6K	2,995K	31.0M	113.42	1.17	9.69
Republican	404	2.2MRUB	$36.0K	647K	8.7M	259.30	3.52	7.36

Although there is no significant difference in the distribution of clicks and impressions between the Ads of the two parties, the Democratic party had significantly higher CTR and lower CPM values. This implies that the Democratic party was more effective amongst the two parties.

However, there was also an active involvement in propagating the Republican Ads as well. This is evident from Table 5 which highlights that the disproportionate high amount spent for the Republican Ads (38.87%) despite their low frequency (17.10%). Moreover, adjudging from the campaign's time-line in Fig. 5 Republican Ads occurred for a longer duration.

Finally, the campaigns of the two parties mostly dealt with conflicting or contradictory ideals (Anti-Islam/Islam, Anti-Immigration/Immigration, All Lives Matter/Black Lives Matter). This strongly suggests desire to sow discord.

We now present the semantic and textual differences between the two parties.

Sentiment Analysis: The compounded sentiment score is significantly lower for the Democratic party since a majority of the Democratic ads pertain to serious topics like police brutality, racial tension, anti-war, etc.

Emotion Analysis: The emotion corresponding to sadness is significantly more pronounced in the Democratic ads due to the above reason.

Structural Analysis: There was no significant difference in the average distribution of words and sentences between the two parties.

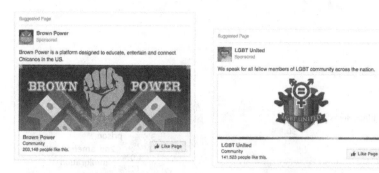

(a) *Democratic* highest clicks (56K) and (b) *Democratic* highest CTR (84.42%)
impressions (968K)

(c) *Democratic* highest cost ($1,200)   (d) *Republican* highest clicks (73K) and
impressions (1.33M)

(e) *Republican* highest CTR(28.16%)   (f) *Republican* highest cost ($5,317)

Fig. 3. Best performing ads for each party

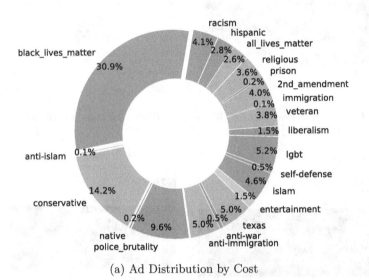

(a) Ad Distribution by Cost

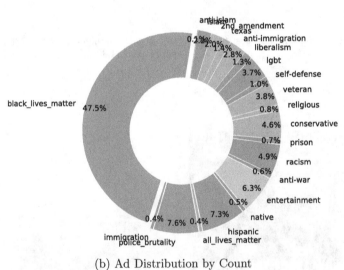

(b) Ad Distribution by Count

Fig. 4. Distribution by impressions, clicks and target age group

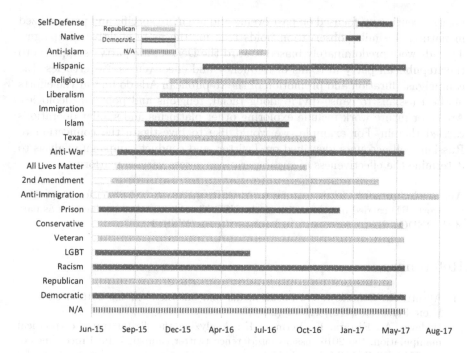

Fig. 5. Timeline of campaigns

POS-TAG Distribution: Surprisingly, plural nouns (both common and proper nouns) occur more frequently in the ads of the Republican party. Adverbs and comparative adjectives are also more prevalent in the Republican ads.

Named Entity Distribution: The fraction of named entities corresponding to *Person* is higher in Democratic ads while those corresponding to *Location* is higher in Republican ads. This happens since the Democratic ads mention the names of victims of racial prejudice like Tamir Rice and Eric Garner. Republican ads of patriotism, veterans, and 2nd Amendment indirectly referenced America (Fig. 3).

LIWC Analysis: The category *We* is more significantly pronounced in the Republican party than the Democratic party which might indicate a closer community or inclusiveness. This may be appealing to the target's sense of belonging (Fig. 4).

7   Conclusions and Future Work

In this paper we characterized the Russian Facebook influence operation that occurred before, during, and after the 2016 US presidential election. We focused on 3,500 ads allegedly purchased by the Russian government, highlighting their features and studying their effectiveness. The most effective ads tend to have less positive sentiment, focused on past events and are more specific and personalized in nature. A similar observation holds true for the more effective campaigns. The ads were predominately biased toward the Democratic party as opposed to the Republican party in terms of frequency and effectiveness. Nevertheless the campaigns' duration and promotion of the Republican Ads do hint at the efforts of the Russians to cause divide along racial, religious and political ideologies. Areas for future work include exploring other platforms and similar operations carried therein. For example, we would like to investigate the connection to Russian troll accounts identified on Twitter, and conduct campaign analysis to determine the effectiveness of such operations across various platforms.

Acknowledgement. EF is grateful to AFOSR (#FA9550-17-1-0327) for supporting this work. RD carried out this work at the USC Viterbi School of Engineering as part of the INDO - U.S. Science and Technology Forum (IUSSTF).

References

1. Allcott, H., Gentzkow, M.: Social media and fake news in the 2016 election. J. Econ. Perspect. **31**(2), 211–36 (2017)
2. Badawy, A., Ferrara, E., Lerman, K.: Analyzing the digital traces of political manipulation: the 2016 russian interference twitter campaign. In: Proceedings of the 2018 IEEE/ACM ASONAM International Conference on Advances in Social Networks Analysis and Mining (2018)
3. Baltas, G.: Determinants of internet advertising effectiveness: an empirical study. Int. J. Market Res. **45**(4), 1–9 (2003)
4. Bessi, A., Ferrara, E.: Social bots distort the 2016 us presidential election online discussion. First Monday **21**(11) (2016)
5. Blei, D.M., Ng, A.Y., Jordan, M.I.: Latent Dirichlet allocation. J. Mach. Learn. Res. **3**, 993–1022 (2003). http://dl.acm.org/citation.cfm?id=944919.944937
6. Blondel, V.D., Guillaume, J.L., Lambiotte, R., Lefebvre, E.: Fast unfolding of communities in large networks. J. Stat. Mech. Theory Exp. **2008**(10), P10008 (2008)
7. Dave, K.S., Varma, V.: Learning the click-through rate for rare/new ads from similar ads. In: Proceedings of the 33rd International ACM SIGIR Conference on Research and Development in Information Retrieval, pp. 897–898. ACM (2010)
8. Davis, C.A., Varol, O., Ferrara, E., Flammini, A., Menczer, F.: BotOrNot: a system to evaluate social bots. In: Proceedings of the 25th International Conference Companion on World Wide Web, pp. 273–274. International World Wide Web Conferences Steering Committee (2016)
9. Del Vicario, M., Zollo, F., Caldarelli, G., Scala, A., Quattrociocchi, W.: Mapping social dynamics on facebook: the brexit debate. Soc. Netw. **50**, 6–16 (2017)
10. Dinh, S., Azeb, T., Fortin, F., Mouheb, D., Debbabi, M.: Spam campaign detection, analysis, and investigation. Digit. Invest. **12**, S12–S21 (2015)
11. Ferrara, E.: Manipulation and abuse on social media. ACM SIGWEB Newsl. (Spring) (2015). 4
12. Ferrara, E.: Disinformation and social bot operations in the run up to the 2017 French presidential election. First Monday **22**(8) (2017)

13. Ferrara, E.: Measuring social spam and the effect of bots on information diffusion in social media. In: Lehmann, S., Ahn, Y.-Y. (eds.) Complex Spreading Phenomena in Social Systems. CSS, pp. 229–255. Springer, Cham (2018). https://doi.org/10.1007/978-3-319-77332-2_13

14. Ferrara, E., Varol, O., Davis, C., Menczer, F., Flammini, A.: The rise of social bots. Commun. ACM **59**(7), 96–104 (2016)

15. Ferrara, E., Varol, O., Menczer, F., Flammini, A.: Detection of promoted social media campaigns. In: ICWSM, pp. 563–566 (2016)

16. Gruzd, A., Jacobson, J., Wellman, B., Mai, P.H.: Social media and society: introduction to the special issue (2017)

17. Howard, P.N., Kollanyi, B.: Bots, # strongerin, and # brexit: computational propaganda during the uk-eu referendum (2016)

18. Hutto, C.J., Gilbert, E.: VADER: a parsimonious rule-based model for sentiment analysis of social media text. In: ICWSM (2014)

19. Kim, Y.M.: The stealth media? Groups and targets behind divisive issue campaigns on Facebook (2018)

20. Kruczkowski, M., Niewiadomska-Szynkiewicz, E., Kozakiewicz, A.: FP-tree and SVM for malicious web campaign detection. In: Nguyen, N.T., Trawiński, B., Kosala, R. (eds.) ACIIDS 2015. LNCS (LNAI), vol. 9012, pp. 193–201. Springer, Cham (2015). https://doi.org/10.1007/978-3-319-15705-4_19

21. Lazer, D., et al.: Life in the network: the coming age of computational social science. Science **323**(5915), 721 (2009)

22. Manning, C.D., Surdeanu, M., Bauer, J., Finkel, J., Bethard, S.J., McClosky,D.: The stanford CoreNLP natural language processing toolkit. In: Association for Computational Linguistics (ACL) System Demonstrations, pp. 55–60 (2014)

23. McCright, A.M., Dunlap, R.E.: Combatting misinformation requires recognizing its types and the factors that facilitate its spread and resonance. J. Appl. Res. Mem. Cogn. **6**(4), 389–396 (2017)

24. Metaxas, P.T., Mustafaraj, E.: Social media and the elections. Science **338**(6106), 472–473 (2012)

25. Mikolov, T., Grave, E., Bojanowski, P., Puhrsch, C., Joulin, A.: Advances in pre-training distributed word representations. In: Proceedings of the International Conference on Language Resources and Evaluation (LREC 2018) (2018)

26. Mohammad, S.M., Turney, P.D.: Crowdsourcing a word-emotion association lexicon **29**(3), 436–465 (2013)

27. Pennebaker, J.W., Francis, M.E., Booth, R.J.: Linguistic Inquiry and Word Count: LIwc 2001. Mahway: Lawrence Erlbaum Associates 71(2001), 2001 (2001)

28. Ratkiewicz, J., Conover, M., Meiss, M.R., Gonçalves, B., Flammini, A., Menczer, F.: Detecting and tracking political abuse in social media. ICWSM **11**, 297–304 (2011)

29. Rehurek, R., Sojka, P.: Software framework for topic modelling with largecorpora. In: In Proceedings of the LREC 2010 Workshop on New Challenges forNLP Frameworks. Citeseer (2010)

30. Ribeiro, F.N., Araújo, M., Gonçalves, P., Gonçalves, M.A., Benevenuto, F.: Sentibench-a benchmark comparison of state-of-the-practice sentiment analysis methods. EPJ. Data Sci. **5**(1), 1–29 (2016)

31. Robinson, H., Wysocka, A., Hand, C.: Internet advertising effectiveness, the effect of design on click-through rates for banner ads. Int. J. Adv. **26**(4), 527–541 (2007). https://doi.org/10.1080/02650487.2007.11073031

32. Rose, S., Engel, D., Cramer, N., Cowley, W.: Automatic keyword extraction from individual documents. In: Text Mining: Applications and Theory, pp. 1–20 (2010)

33. Saher, M., Pathak, J.: Malware and exploit campaign detection system and method, uS Patent App. 14/482,696, 12 March 2015
34. Stella, M., Ferrara, E., De Domenico, M.: Bots sustain and inflate striking opposition in online social systems. arXiv preprint arXiv:1802.07292 (2018)
35. Sunstein, C.R.: # Republic: Divided Democracy in the Age of Social Media. Princeton University Press, Princeton (2018)
36. Toutanova, K., Klein, D., Manning, C.D., Singer, Y.: Feature-rich part-of-speech tagging with a cyclic dependency network. In: Proceedings of the 2003 Conference of the North American Chapter of the Association for Computational Linguistics on Human Language Technology-Volume 1, pp. 173–180 (2003)
37. Varol, O., Ferrara, E., Davis, C., Menczer, F., Flammini, A.: Online human-bot interactions: detection, estimation, and characterization (2017)
38. Varol, O., Ferrara, E., Menczer, F., Flammini, A.: Early detection of promoted campaigns on social media. EPJ Data Sci. 6(1), 13 (2017)
39. Woolley, S.C., Guilbeault, D.R.: Computational propaganda in the united states of America: manufacturing consensus online. Computational Propaganda Research Project, p. 22 (2017)

Semantic Query-Based Patent Summarization System (SQPSS)

K. Girthana[✉] and S. Swamynathan

Department of Information Science and Technology,
Anna University, Chennai 600025, India
keerthi3110@gmail.com, swamyns@annauniv.edu

Abstract. Intellectual property document mainly patents provide exclusive rights to the invention. These technical documents are much valuable to gain insights about the latest trends in the technology and for competitive advancements, R & D management and for future innovations. Since the patent documents are lengthy and contain legal information, it is difficult to process multiple documents manually. This paper proposes Semantic Query-based Summarization System (SQPSS) where the patent search query provided by the patent analyst is enriched with the domain knowledge base, and the retrieved related documents are summarized in an extractive way with the help of Restricted Boltzmann Machine (RBM). Experiments are carried out with search queries from smartphone domain, and the summarization results are evaluated regarding precision, recall and compression rate. The results show an improvement over the existing Open Text Summarizer tool and the summary produced has an average compression rate of around 30%.

Keywords: Domain ontology · Smartphone · Summarization · RBM

1 Introduction

The patent, a legal document with technical information helps in technological development and growth of a country. Every day the number of patent filings and non-patent literature such as Journals, articles, conference proceedings and so on keeps on increasing at a rapid pace [2]. Due to this, the prior-art search process becomes mandatory. Prior-art search mainly helps to gain insights about the latest trends in the technology and also to find a set of prior published or granted patent documents that are relevant to the invention or the patent search query. The existing techniques on prior-art search are either keyword-based or use a domain-independent knowledge base such as WordNet [1,13,14]. The main drawback of the first approach is the patent attorneys mostly draft the patent documents, and so they use more generic terms to claim the scope of the invention. Also, the terminologies are transitory. The problem with a domain-independent knowledge base is it is difficult to find related terminologies for

© Springer Nature Singapore Pte Ltd. 2019
L. Akoglu et al. (Eds.): ICIIT 2018, CCIS 941, pp. 169–179, 2019.
https://doi.org/10.1007/978-981-13-3582-2_13

technical terms. These drawbacks are overcome in the proposed work with the domain ontology for expanding the patent search query.

The prior-art search on a query retrieves thousands of patent documents, and each patent document is around 28–32 pages in length. It is tedious and time-consuming to process (read and understand) these documents manually and to find out whether the invention is patentable or not. So, in this work, a summarization system is proposed that produce the core concepts of a patent document. Generally, the summarization technique [15] can be either extractive or abstractive based on the processing method. As the name implies, extractive summary extracts the salient sentences from a document while the abstractive one identifies core information, rephrases and regenerates the summary. The summary produced through these techniques can be either a single document or multi-documented summary. Also, the summary can be query oriented where the focus is on the topic provided in the query or generic. This Semantic Query-based Summarization System (SQPSS) uses the query oriented extractive technique for summarizing the retrieved prior-art search results.

Document Summarization was done in multiple ways. Most of the existing works are extractive and used statistical and linguistic techniques [18] such as Term Frequency-Inverse Sentence Frequency (TF-ISF), the position of the sentence or paragraph, cue method, word co-occurrence, lexical chains for sentence scoring and ranking. Recent works on text summarization suggest that Deep Learning (DL) achieves impressive results in generating headlines for news articles (Text generation), speech recognition, Machine Translation, and Document Summarization. So, in this Restricted Boltzmann Machine (RBM), a probabilistic neural network is used for producing summaries of the patent documents.

The rest of the paper is organized as follows: Sect. 2 portrays the existing works carried out on prior-art search techniques and document summarization and Sect. 3 discusses the methodology of SQPSS. The Experimental Results carried out as part of this work are detailed in Sect. 4. Finally, Sect. 5 concludes this paper and discusses the possibilities of future enhancements.

2   Related Work

2.1   Prior-Art Search Techniques

Prior-art search involves analyzing the existing knowledge source that need not exist physically. Gaff and Rubinger detail various techniques involved in prior-art searching, along with their need and challenges involved [9]. They also discussed the various ways of prior-art searching. The most common approach for prior-art searching includes patent search query expansion or reformulation. The most common methods for query expansion are through pseudo-relevance feedback (PRF) documents and using some knowledge base or thesaurus. Because of the noise present in the top k documents, PRF is not much suitable for prior-art searching [10,11]. Studies suggest that generic knowledge bases such as WordNet, Wikitionary and Wikipedia categories are used as sources of expansion [1,13]. Because of the lack of domain knowledge, better performance (mainly recall) was not achieved. Mahdabi et al. used an International Patent Classification (IPC) code provided by the patent analyst during application processing for query expansion. They used corresponding code definitions for expansions [14]. This provides for better results in the chemical domain. But the problem is, in some domains the IPC definitions are not detailed enough for all areas in technology.

2.2 Document Summarization

Most of the early work on summarization is generic and uses term frequency to analyze the significance of the sentence [4,12]. Edmundson [8] in addition to the frequency, used positional information and cue words to measure the sentence importance. The documents are scored manually, and the summary is extractive. Some researchers viewed this as a classification problem and applied machine learning algorithms. Features that measure the cohesion among sentences such as word co-occurrence, lexical chain, similarity, and co-reference is used as a sentence scoring mechanism [3,17]. Trappey et al. [20] used text mining techniques such as key phrase recognition and information density approaches to summarize the patent documents through paragraph extraction automatically. Their work focuses on the "Power hand tools" domain. Another group of researchers analyzed the structure and rhetorical relations among the various textual segments [6,16]. They extracted the Rhetorical Structure Theory features to generate the summary. But the construction process is costlier and not scalable. Deep learning is an emerging concept of Machine Learning, and is widely applied in multiple fields of computer science, mainly Natural language processing, Robotics, and Image processing. Recent studies also suggest that summarization using deep learning concepts such as RBM [7], Recurrent Neural Network (RNN) [5] and Convolutional Neural Network (CNN) [21] has reduced issues such as improper sentence formulation, coherence, and redundancy. This work for text summarization uses the unsupervised method to find out the essential features in the patent document and scores the sentences to form a summary.

3   Methodology

The Semantic Query-based Patent Summarization System (SQPSS) was pictorially represented in Fig. 1. The flow of the system is as follows: When the researcher or a patent analyst interested in gaining insights about technology in a domain, then the researcher submits a patent search query that will be automatically expanded semantically with the help of domain ontology. The expanded search query is fed to the Google patent search engine and related patent documents retrieved are filtered based on the International Patent Classification (IPC) code. After pre-processing, the salient sentences are extracted using RBM and submitted to the user as a summary.

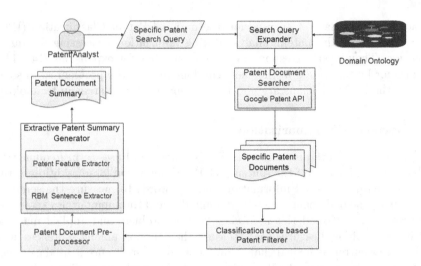

Fig. 1. Flow of semantic query-based patent summarization system

3.1 Smartphone Domain Ontology

The ontology was constructed by considering patent documents and technical specification documents in the smartphone domain. These domain patent documents are crawled using the Google patent search API and fields such as abstract, summary, and description are processed. Using Apache Open NLP, Noun phrases are fetched, and concepts are determined based on Term Frequency-Inverse Document Frequency (TF-IDF) scores. For each concept identified, its related terms such as synonyms, abbreviations, descriptions about it and other similar concepts are retrieved. Relations between the concepts are established, and this ontology acts as a knowledge base for further processing domain related patent search queries. Figure 2 depicts the steps involved in the domain ontology construction and the core concepts derived for the smartphone domain ontology.

3.2 Patent Search Query Expander

This search query expander detects the noun phrases in the query using Part-Of-Speech (POS) Tagger [19] to retrieve related entities. They include subclasses of a concept, instances of the particular concept, synonyms and other similar terms associated with the concept through object properties. Table 1 describes the candidate expansion concepts for each term in the search query. A concepts synonym, abbreviations, and instances are given more preference than other relations to be included for expansion and this work restricts the expanded patent search query length to 30 terms, and a sample patent search query is tabulated in Table 2. From the expanded query, it is known that concepts are extracted only for the terms that occur in the domain ontology.

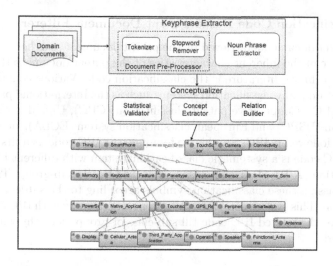

Fig. 2. Smartphone domain ontology

Table 1. Sample query term expansion using smartphone domain ontology

Query: Bluetooth querying device for mobile	
Patent search query term	Candidate expansion terms
Bluetooth	BT, Connectivity, Wireless communication, Bluetooth Antenna and 5 more concepts
Querying device	Bluetooth, LMP, MLME and some more concepts
Mobile	Smartphone

3.3 Patent Document Searcher

The expanded semantic search query when submitted to the patent document searcher, retrieves the several hundreds of related patent documents through the Google patent search API. The Google patent search engine has a collection of patent grants and patent applications from various patent databases across the world. A patent document is structured and composed of multiple fields such as title, abstract, bibliographic information, citations, drawings, descriptions and finally claims. This work considers only title, abstract and description fields of the patent document.

Table 2. Sample patent search query expansion

Patent Search Query	Bluetooth attendance system using mobile application
Expanded Semantic Query	Bluetooth **connectivity BT** attendance system mobile **smartphone** application

3.4   Classification Code Based Patent Document Filterer

All the patent documents retrieved by the patent document searcher need not be relevant. One of the inherent characteristics of the patent document that differs from other technical literature is the classification code. Various country patent offices maintain many classification systems such as the International patent classification (IPC), Cooperative patent classification (CPC), United States patent classification (USPC) and European classification system (ECLA). However, the universally followed is IPC. So, this system uses the IPC code as a first level filter. The IPC code is a systematic classification system with different hierarchies such as section (A-H), class, sub-class, group, and finally sub-group. The patent examiners assign these classification symbols according to the subject matter of the invention. This system compares the patent search query with IPC definitions and retrieves the related IPC code till sub-class or group hierarchy whichever is more related.

3.5   Patent Document Pre-processor

Multiple versions of the same patent document are available on the web. In some versions of a patent document, some of the fields are left blank intentionally. If such documents are retrieved, the fields are updated with the final version of the same patent document and a single copy is maintained for further processing. Pre-processing involves removing meaningless and irrelevant words. In addition to standard English stop words, patent specific stop words such as "invention", "description", "embodiment", "present" and so on are also removed.

3.6   Extractive Patent Summary Generator

Patent summary generation is a two-step process, namely patent feature extraction and salient sentence extraction.

Patent Feature Extraction. It involves extracting features such as title and search query similarity, sentence field position, and Term frequency – Inverse field frequency (TF-IFF) and are detailed below:

- **Title and search query similarity** is determined based on the number of common occurrences of words between the title, search query, and the sentences. The variations among the various field lengths are overcome by normalizing the similarity score with the logarithmic values of their length.
- The **position of the sentence** within a field also plays a more significant role. It is based on the assumption that the sentence at the beginning will be more like introductory, sentences towards the end will move towards a conclusion, and the scores are assigned based on Eq. 1.

$$SF_i = \begin{cases} 0.25, & \text{if } sen_i \text{ is at the beginning} \\ 0.75, & \text{if } sen_i \text{ is at the end} \\ max[i^{-1}, (m-i+1)^{-1}], & \text{otherwise} \end{cases} \qquad (1)$$

Where i is the currently processed sentence index and m is the total number of sentences considered for a document.

- **Term Frequency and Inverse Field frequency** is computed for the noun phrases of the form $(<JJ>^*<NN.^*> + <IN>)?<JJ>^*<NN.^*>+$ Where JJ is an Adjective, NN. represents Noun phrases and IN is a preposition or subordinating conjunction. The score of the sentence based on TF-IFF is the summation of all noun phrases TF-IFF scores and the same is represented in Eq. 2.

$$TF - IFF_i = \sum_{tesen_i} TF - IFF_{t,f} \tag{2}$$

RBM Sentence Extractor. The features extracted in the feature extraction step forms the sentence-feature matrix, and the salient sentences are extracted using RBM. RBM has three perceptrons at the visible layer, and the hidden layer is composed of two perceptrons. The system is trained using contrastive divergence with a learning rate of 0.1. Each sentence feature vector is passed to the hidden layer along with the learned weights and biases. Once learning is completed, it outputs binary values indicating whether the sentence needs to be included for a summary or not. For each document, the sentences, whose RBM output is one, are included in the salient sentences set and are provided as a summary to the user. Table 3 portrays the summary generated using RBM for a sample input text. The input text describes the method of taking attendance using wireless technology and the extractive summary generated clearly produces the gist of the input text.

Table 3. Generated extractive summary for sample text

Input text	Extractive summary
The invention discloses an attendance method and system based on wireless interconnection technology. The method comprises the following steps: an attendance machine sends an attendance request to terminal equipment, wherein the attendance machine is connected with the terminal equipment by the wireless interconnection technology; the terminal equipment sends identification information to the attendance machine; the attendance machine sends the received identification information to an attendance server; and the attendance server authenticates identity of the received identification information according to the pre-stored authentication information and sends the authentication result to the attendance machine. The invention can improve the attendance efficiency and save the attendance cost	The invention discloses an attendance method and system based on wireless interconnection technology. An attendance machine sends an attendance request to terminal equipment. The terminal equipment sends identification information to the attendance machine. The invention can improve the attendance efficiency and save the attendance cost

4   Experimental Results

Experiments are conducted to evaluate the performance of the proposed summarization system regarding precision, recall and compression rate. The evaluated results are compared to an existing Open text summarizer[1] tool. Precision and recall are given by Eqs. 3 and 4 where $Summ_{Act}$ describes actual summary provided by human summarizers and $Summ_{pred}$ represents the summary of the proposed system. Compression rate is the amount of information retained in the generated summary when compared to the original document. These experiments were conducted by issuing a generated semantic patent search query to the Google patent search engine through Google patent API. The evaluation metrics are calculated by retrieving top 10 summarized documents for each patent query. The results of each module for a sample query "Bluetooth attendance system using mobile application" are detailed in the following subsections.

$$precision = \frac{|Summ_{Act} \cap Summ_{pred}|}{|Summ_{pred}|} \tag{3}$$

$$recall = \frac{|Summ_{Act} \cap Summ_{pred}|}{|Summ_{Act}|} \tag{4}$$

4.1   Classification Based Patent Document Filterer

This expanded semantic search query, when fed to the search engine, retrieves 829 patent documents and these documents are filtered with the IPC code G07C, H04L and H04W4 as these codes are related to "Attendance Registers", "Transmission of Digital Information through Telephonic communication" and "Services for Wireless communication Networks" respectively. Some of the other codes that were present in the majority in the retrieved patent document set were G06Q30 (Commerce/Shopping), G06Q50 (Utilities/Tourism) and H04W84 (Topologies). The same is tabulated in Table 4 with the number of documents in each classification code.

Table 4. Patent Document Filtering for a sample query

Search query	Bluetooth beacon attendance management
Number of patent documents initially retrieved	829
IPC code based Filtering	G07C (53), H04L (115), H04W4 (100)
Other IPC codes in the retrieved document set	G06Q30 (52), G06Q50 (41), G06F17 (16), H04W84 (24)
Number of patent documents after filtering	238

---

[1] https://www.splitbrain.org/services/ots.

4.2 Extractive Summary Generator

To generate the summary, first, the features of the patent document are extracted. The extracted features of a document form the sentence-feature matrix. The sentence feature vector, when fed to the RBM, outputs the binary value for the sentences to be included in the summary. For this sample query, Table 5 depicts the summary statistics along with the evaluation metrics for top 10 retrieved patent documents after filtering. It can be observed that the average compression rate of the summaries obtained using the proposed approach is around 30%. This proposed work is compared with the existing open text summarizer tool, and the summaries are retrieved by initializing the compression rate to 30%. The summaries produced through both approaches are evaluated in terms of precision and recall. The results show that SQPSS summarises better when compared to Open Text Summarizer.

Table 5. SQPSS evaluation metrics

Patent title	Total no. of sentences			SQPSS		Open text summariser	
	Before extraction	After extraction	Comp. rate	Precision	Recall	Precision	Recall
CN101916462A	101	78	23%	0.780	0.717	0.79	0.5
CN103793833B	53	38	28%	0.675	0.831	0.53	0.54
CN104392501A	92	64	30%	0.732	0.811	0.61	0.59
CN105867321B	90	67	26%	0.605	0.732	0.49	0.55
CN106023332A	97	72	26%	0.676	0.740	0.51	0.58
CN104299276A	34	23	32%	0.613	0.678	0.46	0.571
CN106382932A	89	65	27%	0.710	0.730	0.51	0.59
CN107147993A	81	69	15%	0.595	0.618	0.43	0.49
JP2013500538A	86	65	24%	0.664	0.791	0.567	0.581
US9734643B2	437	300	31%	0.735	0.795	0.41	0.48

5 Conclusion

This paper presented an experimental study of semantic query-based patent summarization system (SQPSS) using Domain ontology and Deep Learning algorithm RBM. The domain ontology is used to enrich the patent search query further, thereby providing more related documents to the patent analyst. Summarization is the process of producing the core content by reducing the less critical information. Since the deep learning algorithm mimics the human brain, it was used in this system. The features extracted from the retrieved patent document set for a search query form the sentence-feature matrix and are sent for processing in various layers of RBM. Generated summaries are evaluated regarding precision, recall, and compression rate for different search queries and the same is explained with an example query and different documents retrieved using the query. The SQPSS when compared to the existing summarization tool, open text summariser shows better results such as precision (0.73), recall (0.767) and average compression rate (30%). It was observed during experimentation, that SQPSS misses some of the salient sentences found to be relevant by human summarisers. This may be because it focuses more on the frequency and similarity measures. To overcome this, the futuristic enhancements to the SQPSS involve including the semantics for feature extraction and upgrading the performance with stacked RBM. Also, the domain ontology can be further extended to other domains.

Acknowledgments. This research is sponsored by Visveshvaraya PhD Scheme for Electronics & IT Proceedings No. 3408/PD6/DeitY/2015.

References

1. Al-Shboul, B., Myaeng, S.H.: Wikipedia-based query phrase expansion in patent class search. Inf. Retrieval **17**(5–6), 430–451 (2014)
2. Alberts, D., et al.: Introduction to patent searching. In: Lupu, M., Mayer, K., Kando, N., Trippe, A. (eds.) Current Challenges in Patent Information Retrieval. TIRS, vol. 37, pp. 3–45. Springer, Heidelberg (2017). https://doi.org/10.1007/978-3-662-53817-3_1
3. Azzam, S., Humphreys, K., Gaizauskas, R.: Using coreference chains for text summarization. In: Proceedings of the Workshop on Conference and its Applications, pp. 77–84. Association for Computational Linguistics (1999)
4. Berger, A., Mittal, V.O.: Query-relevant summarization using FAQs. In: Proceedings of the 38th Annual Meeting on Association for Computational Linguistics, pp. 294–301. Association for Computational Linguistics (2000)
5. Cao, Z., Wei, F., Dong, L., Li, S., Zhou, M.: Ranking with recursive neural networks and its application to multi-document summarization. In: AAAI, pp. 2153–2159 (2015)
6. Chuang, W.T., Yang, J.: Extracting sentence segments for text summarization: a machine learning approach. In: Proceedings of the 23rd Annual International ACM SIGIR Conference on Research and Development in Information Retrieval, pp. 152–159. ACM (2000)
7. Duraiswamy, K.: An approach for text summarization using deep learning algorithm (2014)
8. Edmundson, H.P.: New methods in automatic extracting. J. ACM (JACM) **16**(2), 264–285 (1969)
9. Gaff, B.M., Rubinger, B.: The significance of prior art. Computer **8**, 9–11 (2014)
10. Ganguly, D., Leveling, J., Magdy, W., Jones, G.J.: Patent query reduction using pseudo relevance feedback. In: Proceedings of the 20th ACM International Conference on Information and Knowledge Management, pp. 1953–1956. ACM (2011)
11. Lopez, P., Romary, L.: Experiments with citation mining and key-term extraction for prior art search. In: CLEF 2010-Conference on Multilingual and Multimodal Information Access Evaluation (2010)
12. Luhn, H.P.: The automatic creation of literature abstracts. IBM J. Res. Dev. **2**(2), 159–165 (1958)
13. Magdy, W., Jones, G.J.: Applying the kiss principle for the CLEF-IP 2010 prior art candidate patent search task (2010)
14. Mahdabi, P., Crestani, F.: Patent query formulation by synthesizing multiple sources of relevance evidence. ACM Trans. Inf. Syst. (TOIS) **32**(4), 16 (2014)
15. Mani, I.: Automatic Summarization, p. 3. John Benjamins Publishing, Amsterdam (2001)

16. Marcu, D.: The rhetorical parsing of natural language texts. In: Proceedings of the 35th Annual Meeting of the Association for Computational Linguistics and Eighth Conference of the European Chapter of the Association for Computational Linguistics, pp. 96–103. Association for Computational Linguistics (1997)

17. Radev, D.R., Jing, H., Budzikowska, M.: Centroid-based summarization of multiple documents: sentence extraction, utility-based evaluation, and user studies. In: Proceedings of the 2000 NAACL-ANLP Workshop on Automatic summarization, vol. 4. pp. 21–30. Association for Computational Linguistics (2000)

18. Torres-Moreno, J.M.: Automatic Text Summarization. Wiley, New York (2014)

19. Toutanova, K., Klein, D., Manning, C.D., Singer, Y.: Feature-rich part-of-speech tagging with a cyclic dependency network. In: Proceedings of the 2003 Conference of the North American Chapter of the Association for Computational Linguistics on Human Language Technology, vol. 1, pp. 173–180. Association for Computational Linguistics (2003)

20. Trappey, A.J., Trappey, C.V.: An R&D knowledge management method for patent document summarization. Industr. Manage. Data Syst. **108**(2), 245–257 (2008)

21. Zhang, Y., Er, M.J., Zhao, R., Pratama, M.: Multiview convolutional neural networks for multidocument extractive summarization. IEEE Trans. Cybern. **47**(10), 3230–3242 (2017)

Proposed Strategy for Allergy Prediction Based on Weather Forecasting and Social Media Analysis

Sugandha Sharma[✉], Anmol Sachan[✉], and Harneet Singh[✉]

Chandigarh University, Chandigarh, India
sugandhasharma046@gmail.com,
anmolsachan1997@gmail.com, Harneetsgh98@gmail.com

Abstract. Allergies have recently become the most prevailing problem after all the technological and industrial development. Human body system has reached the limit where it cannot adapt to such high degrees of pollution and even the natural things at times. Understanding the responsibility of being the most superior race on this planet, humans have implemented technology and its various aspects for the betterment of other species whether it be plant species or animal species. Humans are trying to move towards advancements and taking other living beings with them. At the moment humans have created problems for themselves by destroying the natural flora and fauna. This is debatable that humans did it for everyone or for their selfish sake but no one can deny that damage has been done. The consequences being diseases and allergies all such health issues have increased with time. Now, something had to be done to come over it after all humans are most superior. Again technology came to the rescue and humans implemented systems that can predict weather and climatic changes so that a person can be ready with all the precautionary measures. This paper discusses allergy forecast system. Allergy prediction system will work on a similar principle of data collection and using that data for predicting possible health issues that might occur as a result of the climatic condition in that area. The system collects information from weather forecast system to show possible health problems based on climate and also uses data from social media (i.e. social media analysis) to predict possible problems based on updated form different regions.

Keywords: Allergies · Environment · Health · Diseases · Prediction
Social media analysis

1 Introduction

[12] Health issues and weather is something that cannot be predicted to a 100% accuracy but, all that could be done is improving the accuracy and this is what researchers and people into development have been trying to do. Protocol with everything in this world which is new is same. First, the prototype is developed then it is put to an actual test. Obviously, a technology is always not that great in the initial days but with time and money that is invested into it definitely gives some output. There is some relationship between one's health and the climate he/she lives in. There used to be days when people

L. Akoglu et al. (Eds.): ICIIT 2018, CCIS 941, pp. 180–189, 2019.
https://doi.org/10.1007/978-981-13-3582-2_14

were not affected by minute things like dust and weather change but now, the immunity levels and the quality of the environment has decreased. The human body has adapted to the climate change but isn't immune to the problems that come with the climatic condition. Allergies are easily triggered in the body with the least amount of allergy-causing element in the surrounding. It has become really important for a human being to be aware of what is happening in surrounding and inside the body (the Primary reason behind the success of fitness bands and smart watches). It is important because precautionary measures could be only taken for a particular thing only if one is aware of its occurrence. General health issues are caused basically because of 2 reasons that are:

- Environmental factor: This is basically the climate and the weather of an area. The seasons have a characteristic property that triggers some part of the body that responds to climatic change. For example, human skin tans that are because of the production of excess melanin, which is a response of skin to excessive exposure to sunlight. This is a protective measure that a human body takes itself and is not that bad but if the exposure increases the skin might get damaged.
- Social factor: Not all Diseases are caused due to environment some are exchanged between humans only are human bodies and surroundings also act as carriers for the same. For example, Chickenpox would not spread if the infected person is isolated; it spreads only when someone comes in direct contact with or surrounding of the patient. Diseases that are inherited from someone in the family also come under this category.

Once again the probability of one getting infected with a certain health issue depends upon how the body will react to the change.

This paper discusses the role of technology in spreading awareness among masses about any climatic change or social issue that can cause health problems.

2 Allergy Prediction Based on Climate

[13] Most of the allergies are based on environment or are connected to certain elements in the environment that triggers allergy or histamines in the body. The body treats those elements as foreign harmful substances and produces substances that counter them. The act of producing antibodies for foreign substances is acceptable but those things cause some other problems. For example, a person who has respiratory problems will have a problem with pollen grains. Pollen grains in simplest words are gametes of flowers. These gametes get into the atmosphere with air.

Bodies that respond aggressively to pollens will trigger histamines to a greater extent, which might lead to excessive mucus production, saliva production or swelling in the nasal passage or windpipe which leads to obstruction in breathing.

[3] Body's response to pollen is justified; all these things will stop pollens from getting into the respiratory system or broadly speaking getting into the body. But this is creating new problems for the person.

Similarly, in the rainy season with the first rainfall a smell of soil rises up that stimulates histamine in the body and as a result, the body shows some allergic effects. A person with a problem like this will face health issues in areas where there is rainfall.

So, climatic allergies or diseases are based on the environmental factors like temperature, water or air quality, the season of the year etc.

3   Allergy Predictions Based on Social Media Analysis

Social media analysis is a great tool that companies have been using to predict the consumer requirements or even knowing what the majority of consumer dream about or expect from a product/service. This was possible because of the great tendency of uploading/updating everything on social media websites (Fig. 1).

The social media analysis can be broken down into 3 major components:

1. Monitoring
2. Measuring and analysis
3. Reporting

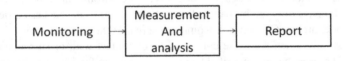

Fig. 1. Brief workflow of social media analysis

[2] First things first, in the monitoring phase the analysis system or the analyst fetches data from multiple sources online like Twitter & Facebook etc. (Twitter being the most famous one). This stage is further divided into two stages i.e. User Interaction and Qualitative research. In these two phases, the analysis system/analyst fetches data from selected sources and only the useful data from the entire system is taken so as to minimize the work. For example, an analyst wants to do an analysis of Effect of Poliovirus then they will first have to target a platform they want to analyze and then filter all the information regarding Polio Virus from the rest of the data on that platform. This stage also includes filtering the type of data to be fetched like survey reports, posts, #tags etc.

[6] Secondly, the entire selected data is now gathered and is set for an analysis. In the second phase of analysis, the data is looked out for patterns and terms that have to be analyzed just to get a more focused data set. The biggest problem with social media analysis understands the sentiment and notion of anything online. This is where content analysis comes into the role and using AI and ML analysts have been able to get the sensible meaning of that particular thing online (Figs. 2 and 3).

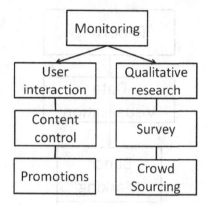

Fig. 2. Components of monitoring stage of social media analysis for allergy prediction

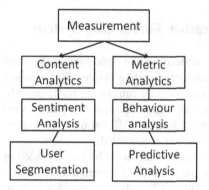

Fig. 3. Components of monitoring stages of social media analysis for allergy prediction

[9] Finally, in the Reporting stage, the result of the analysis is generated and that can be in multiple forms including text, charts, and graphs etc. This is achieved by data visualization which converts the processed data into a more simple graphical representation out of which making conclusions in easy (Fig. 4).

[4] Implementation of social media analysis for allergy prediction is not limited to the steps mentioned, there is always scope for improvement and furthermore, layers could be added or removed to make the system better.

[11] So, Predictions of allergy using social media in a narrower term means that getting the information about the current allergies and diseases that prevail in an area and using that data to warn about some allergy which a person might get infected with it gets in the radius of the infection.

"There has to be something in this if all Blue Chip companies have a dedicated team for doing social media analysis".

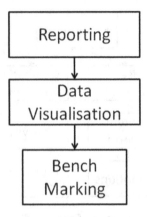

Fig. 4. Final reporting stage

4 Accuracy of Weather Forecasting System

[7] Learning takes time and learning the right thing in depth takes even more time but, it's not true about the weather forecasting system. The accuracy of the weather forecast cannot be 100% because the weather can change really quickly. The older or historical data is only used to know the weather pattern of a place. Weather forecasting systems use data from multiple input sources. All that data is combined together and a possibility of changes in the weather is predicted.

Changes in weather cannot be predicted by just considering one or two entities, for a greater accuracy multiple aspects of environment needs to be considered. Like Humidity, Wind speed, Tides, UV radiations etc. All such data is collected through various sensors and systems like for tides buoys are thrown in the sea and then changes in the sea are observed. Changes in the sea can cause multiple climatic changes like rain & droughts. The interesting fact here is that tides are said to be controlled by the moon so moon could also be put to study to predict weather on Earth. Possibilities are limitless just the systems need to be improved (Figs. 5 and 6).

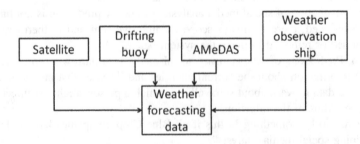

Fig. 5. Sample input sources of weather forecasting system

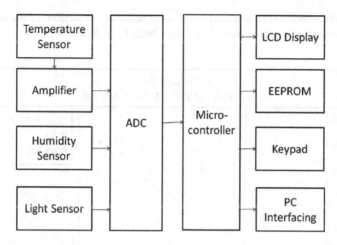

Fig. 6. A general workflow of weather forecast system

There are predictions for a day, a week, and a fortnight and even for a month but the most precise one is prediction done for the next day (Fig. 7).

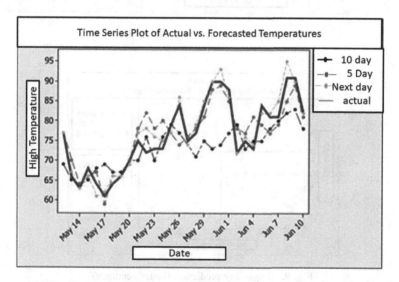

Fig. 7. Sample weather prediction data for 30 days actual vs forecasted

[8] This is again because of the dynamicity of the climate. Our systems are cannot predict the change in Tides or wind which play a crucial role in determining the weather. The changes or movements in winds and tides are not traceable to 100% precision when it comes to predicting for 5–7 days or more. When it comes down to predicting the next day weather, the movement patterns become more track able

changes can be predicted which lead to the most accurate prediction for the next day at least more accurate than one done 5–10 days back Figs. 8 and 9.

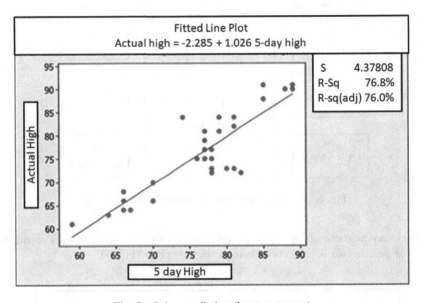

Fig. 8. 5-day prediction (lesser accuracy)

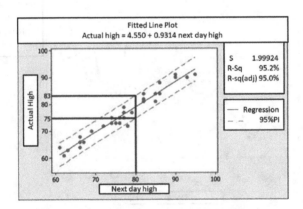

Fig. 9. Single day prediction (better accuracy)

The predictions made for more than a day are just tentative weather conditions based on the previous and present-day climatic condition. Those tentative ideas are only useful in getting an idea for the future since human beings plan out things and have that urge to know the future as well.

5 Accuracy of Social Media Analysis

[1] Weather forecast can get wrong because of the dynamic nature of weather similarly, social media stats cannot be 100% accurate because of dynamicity of humans. The problem with social media analysis is that people do not post things straight away they like emotions to it and understanding that those sentiments and getting the real meaning out of the words from a post are difficult for a machine. Datasift world's largest company to analyze data on Twitter has quoted that only 70% accuracy has been achieved i.e. 70% sentimental analysis are correct. Their study says that even with humans 85% accuracy can be achieved in sentiment analysis over social media (Table 1).

Table 1. Comparative study of social media analysis and weather forecast system

Parameters	Social media analysis	Weather forecast system
Source	Collection by users	Through satellites
Accuracy	50%	70%
Predictive	no	yes
Data collection	By users	Through companies
No. of sources	Crowdsourcing	40–50
Responsible	No one	companies

6 Proposed Strategy to Predict Allergy

Just like any other problem humans are continuously trying to find solutions to health issues independent of it being from any sector. Now, as discussed earlier no one can predict weather and health with 100% accuracy but still, there are services that can give an indication based on the current climatic or social status.

[10] In the health sector the probabilities are not countable, there has to be some sold base which if not 100% accurate then at least be some precise enough to get positive results. In order to increase the accuracy of allergy prediction, a combination of both the services i.e. weather forecast system and social media analysis will be clubbed together.

The weather system will get the possible health problems based on the climatic or environmental condition which are more likely to occur. Their accuracy is not that good for 10–15 days prior predictions. Services that consider climate work like this:

- The system is loaded with a database that contains stats from previous records or findings from experiments which tell about at what level of certain environmental condition an allergy will be triggered.
- The current weather status of an area is fetched for observations.
- The weather condition is matched with the database and predictions are made that suggest what will be the possible allergies that will be associated with that particular climatic condition.

Services that take social media or other media are into consideration take feeds from different users. Basically, it utilizes social media analysis. Users update their status or take a survey or make searches for a particular disease/allergy/symptom online and all that is summed up and the most common one is selected as final. For example, A certain number of people report Delhi to be infected with Dengue on social media then the updates and alerts will be released for people living or going to Delhi. Social media analysis will not only tell about allergies but it can also warn about natural mishappenings and other communicable diseases.

[5] Such systems are not much dependable but there is some dependability that is why they are used to such an extent.

The accuracy of outcomes can go really up and actual people are feeding information to the systems based on the current happenings which are more trustable in a long-term process. Social media analysis of areas will also help in pinpointing the location and will bring up exact health issues faced by an area or possibly could be faced.

The paper suggests a strategy in which both the technologies (weather forecasting and Social media analysis) will work hand in hand and a cumulative suggestion will be generated for the user. In this way, the user gets the allergy prediction on the basis of weather and big data related to the health of that area plus any airborne or other diseases that are spread in a region.

The future scope of the proposed approach is limitless as the kind of information we can get from social media analysis has no bounds and kind of importance SMA is getting, there is no doubt that it will end up becoming a great gimmick for mankind.

7 Discussions and Outcomes

There are already systems ready for sickness prediction or sickness forecast but all of them are not so dependable due to the fact that humans are their own enemies. To get better outputs out of the systems working for allergy and disease prediction a combination of both weather forecast and social media analysis can be used under one roof.

The naturally occurring allergies and diseases (i.e. the ones that occur due to some climatic or environmental change) can be forecasted using the weather system and things at ground reality (i.e. the exact happening and live conditions) can be fetched using the social media analysis.

Combining these two can also help in controlling the pollution levels by indicating the actual pollution level and then linking it to social media and bringing up the concerns of people and problems faced by them on a wider platform. The increase in the number of sample size (observations) will give better accuracy. For the future, there can be more systems that can be combined to this combination just to get better results.

References

1. Asur, S., Bernardo, A.H.: Predicting the future of social media. In: Proceedings of the 2010 IEEE/WIC/ACM International Conference on Web Intelligence and Intelligent Agent Technology WI-IAT 2010, vol. 01 (2010)
2. Anandhan, A., Shuib, L., Ismail, M.A.: Social media recommender systems: review and open research issues. IEEE Access **6**, 15608–15628 (2018)
3. Alun, P., et al.: Sentinel: a codesigned platform for semantic enrichment of social media streams. IEEE Trans. Comput. Soc. Syst. **5**(1), 118–131 (2018)
4. Zeng, D., Chen, H., Lusch, R.F., Li, S.H.: Social media analytics and intelligence. IEEE Intell. Syst. **25**(6), 13–16 (2010)
5. Tang, L., Liu, H.: Community detection and mining in social media. In: Synthesis Lectures on Data Mining and Knowledge Discovery
6. Kane, G.C., Alavi, M., Labianca, G.J., Borgatti, S.P.: What's different about social media networks? a framework and research agenda Year. **38** (2014)
7. Lorenc, A.C.: Analysis methods for numerical weather prediction. Q. J. R. Meteorol. Soc. **120**, 1367–1387 (1986)
8. Kalnay, M.K., Wayman, E.B.: Global numerical weather prediction at the national meteorological center. Bull. Am. Meteorol. Soc. Eugenia (1990)
9. Leith, C.E.: Objective methods for weather prediction. Ann. Rev. Fluid Mech. **10**, 107–128 (1978)
10. Kuo, Y.-H., Sergey, S., Richard, A.A., Francois, V.: Assimilation of GPS radio occultation data for numerical weather prediction. In: Terrestrial Atmospheric and Oceanic Sciences (2000)
11. Klinger, U.: Mastering the art of social media. Inf. Commun. Soc. **16**(5), 717–736 (2013)
12. Ho, W.-C., et al.: Air pollution, weather, and associated risk factors related to asthma prevalence and attack rate. Environ. Res. **104**(3), 402–409 (2007)
13. Cawthorn, C.: Weather as a strategic element in demand chain planning. J. Bus. Forecast. Methods Syst. **17**(3), 18 (1998)

Wind Characteristics and Weibull Parameter Analysis to Predict Wind Power Potential Along the South-East Coastline of Tamil Nadu

P. S. Maran[1(✉)], P. M. Velumurugan[2], and B. Prabhu Dass Batvari[2]

[1] Department of Computer Science and Engineering,
Sathyabama Institute of Science and Technology, Chennai, Tamil Nadu, India
psmaran@sathyabama.ac.in
[2] Centre for Earth and Atmospheric Sciences, Sathyabama Institute of Science
and Technology, Chennai, Tamil Nadu, India

Abstract. The main objective of this paper was to analyze the statistical wind data obtained from the measurements of Automatic Weather Station (AWS), India Meteorological Data (IMD) located in the south-east coastline of Tamil Nadu. In this study the Wind Power Density (WPD) estimation output for a small scale wind power from the surface wind data was analyzed using Weibull parameters and maximum likelihood and least square methods. The Wind Speed and Wind Direction frequency distributions from the year 2009 to 2013 were analyzed. The Weibull parameters are determined based on the wind distribution statistics calculated from the measured data. The Wind Power Density out with the actual data were compared and it is shown that the Weibull representative data estimate the wind energy output very accurately.

Keywords: Wind energy · Weibull distribution · Automatic weather station

1 Introduction

Wind energy is one of the most important sources of renewable energy. Quantifying wind speed data is essential in investigating the availability of wind energy. Wind Characterizing at a specific location or any other area is very important. This process is highly complex due to the nature of wind, which doesn't depend on the statistical distributions. The most suitable areas in Tamil Nadu for wind power generation are the districts of Tirunelveli, Thoothukudi, Kanyakumari, Theni, Coimbatore, and Dindigul. Along the south-eastern coastline of Tamil Nadu, there are no valleys and mountains. Besides, many of these regions are far from the sea. Also, the northerly winds are not as strong as the southerly winds. With regard to the areas surrounding the Indian Ocean, Kanyakumari district is especially suitable for wind power harvesting.

In deciding on a suitable site with wind energy potential, crucial parameters for a wind turbine are the size and shape of the blade, wind direction, the total capacity, etc. A probability study of the site is also required with basic properties such as wind behaviour, wind availability, wind probability distribution and continuity of wind speed in the regions. In this research, the wind quality in the south-eastern coastal

© Springer Nature Singapore Pte Ltd. 2019
L. Akoglu et al. (Eds.): ICIIT 2018, CCIS 941, pp. 190–199, 2019.
https://doi.org/10.1007/978-981-13-3582-2_15

regions of Tamil Nadu such as Chennai, Kalpakkam, Pondicherry, Karaikal, Nagap-
attinam and Thoothukudi were considered. A feasibility study of the site was also
needed with basic properties including wind behaviour, wind availability, wind
probability distribution and continuity of wind speed in the region. Prior to the erection
of the windmills at these sites, these primary properties were studied. This information
is required for investors in any wind energy project. Also, Weibull distributions were
carried out for wind speed on an hourly, weekly and monthly basis along with wind
direction.

Generally, the calculated wind speed data is used in the evaluation of wind energy
potential. Initially, the hourly wind velocity and direction information are observed and
monitored at any location. The frequency distribution and probability distribution
modelling is developed with the interpretation of the possibilities. The wind velocity
details are normally arranged in a time series manner for the frequency distribution
model. The main tools are wind speed distributions and it is denoted by mathematicians
in literature. In scientific research, the Weibull distribution is well utilized to analyze
important functions and parameter estimations.

2 Area of Study

This research work was carried out by collecting wind parameter values from Automatic Weather Stations (AWS) and Meteorological tower stations located in the south-
eastern coastline of Tamil Nadu. The Automatic Weather Stations (AWS) were laid
down by the Indian Meteorological Department (IMD) and the 50 m Meteorological
tower stations were set up by Sathyabama University, Chennai and Indira Gandhi
Centre for Atomic Research (IGCAR), Kalpakkam. The anemometer height of Automatic Weather Stations was above 10 m from the surface. The AWS locations and the
geographical coordinates with their altitude are shown in Table 1. There are no
obstacles in and around the wind speed measuring stations. In summation, we can say
that there were no obstructions in these positions which could make an impingement on
wind speed and twist management. The wind parameter meta data include hourly wind
velocity and wind directional status for the period 2009 to 2013.

Table 1. Locations of stations

Sl. No	Site Location	Latitude (North)	Longitude (East)	Altitude (m)	Height of anemometer (m)
1	Ennore port	13°.2'	80°.3'	5	10
2	Nungambakkam	13° .7'	80° .2'	6	10
3	Sathyabama University	12° .9'	80° .0'	7	10
4	Anupuram, kalpakkam	12° .3'	80° .1'	5	10
5	Karaikal	10° .9'	79° .8'	7	10

(*continued*)

Table 1. (*continued*)

Sl. No	Site Location	Latitude (North)	Longitude (East)	Altitude (m)	Height of anemometer (m)
6	Chidambaram	11° .4'	79° .7'	4	10
7	Adhiramapattinam	10° .4'	79° .4'	4.3	10
8	Thiruchendur	8° .5'	79° .2'	4	10
9	Thoothukudi port	8° .8'	78° .2'	0.64	10
10	Neyyoor	8°.2'	77° .3'	47	10

3 Methodology

3.1 Weibull Parameter Analysis

It is well acknowledged that some hard work is needed to build a sufficient statistical wind speed frequency distribution model for predicting wind energy potential output. For developing wind energy applications, the Weibull distribution model is broadly accepted and the model always gives excellent wind parameter statistics. The Weibull distribution model described two parameters for wind speed variability studies. The Weibull distribution is a standard approach to wind load probability study. It can be utilized with a huge collection of weather parameters. Wind data for the years 2009–2013 were collected from the weather stations. The Weibull shape parameter (k) and the Weibull scale parameter (c) were calculated according to this data using windographer software.

The Weibull distribution with two parameters is exhibited as follows;

$$f_W(v) = \left(\frac{k}{c}\right)\left(\frac{v}{c}\right)^{k-1} \exp\left[-\left(\frac{v}{c}\right)^k\right] \tag{1}$$

Where $f_w(v)$ is the probability of wind velocity(v), the weibull shape parameter (k) and the weibull scale parameter(c). The weibull distribution of cumulative probability function is

$$f_W(v) = 1 - \exp\left[-\left(\frac{v}{c}\right)^k\right] \tag{2}$$

Readily available methods are there to calculate k and c values. Some important methods are maximum likelihood, least square, Average wind velocity and standard deviation (Gokcek et al. 2007). Here, the two values of k and c are calibrated with average wind velocity and standard deviation. As per the process, such parameters are expressed in Eqs. (3) and (4) respectively (Gokcek et al. 2007).

$$k = \left(\frac{\sigma}{v_m}\right)^{-1.086} \quad (1 \leq k \leq 10) \tag{3}$$

$$c = \frac{v_m}{\Gamma(1+1)/k} \tag{4}$$

The average wind speed V_m is derived using Eq. (5). The standard deviation σ is derived using Eq. (6). The Gamma function is $\Gamma()$ is derived for any y value written in Eq. (7) (Gokcek et al., 2007).

$$v_m = \frac{1}{n}\left[\sum_{i=1}^{n} v_i\right] \tag{5}$$

$$\sigma = \left[\frac{1}{n-1}\sum_{i=1}^{n}(v_i - v_m)^2\right]^{1/2} \tag{6}$$

$$\Gamma(y) = \int_0^{\infty} \exp(-x)x^{y-1}dx \tag{7}$$

Where n denotes the no.of hours, years, months and seasons.

The values of k and c were strongly related to the average wind velocity (v_m) (Akpinar 2004)

$$v_m = c\Gamma\left(1 + \frac{1}{k}\right) \tag{8}$$

The power of wind velocity v in the blade swept area A is given by (Celik 2003)

$$P(v) = \frac{1}{2}\rho A v^3 \tag{9}$$

In any probability density function, the wind power density in a particular spot per unit can be given as (Akpinar 2004)

$$P(v) = \frac{1}{2}\rho \int_0^{\infty} v^3 f(v)dv \tag{10}$$

The air density r = 1.225 kg/m3. Based on Weibull distribution, the wind power density per unit is calculated as

$$P_W = \frac{1}{2}\rho c^3 \Gamma\left(1 + \frac{3}{k}\right) \tag{11}$$

$$v_m = c\Gamma\left(1 + \frac{1}{k}\right) \tag{12}$$

The power of wind velocity v in the blade swept area A is given by (Celik 2003)

$$P(v) = \frac{1}{2}\rho A v^3 \tag{13}$$

In any probability density function, the wind power density in a particular spot per unit can be given as (Akpinar 2004)

$$P(v) = \frac{1}{2}\rho \int\limits_0^\infty v^3 f(v) dv \tag{14}$$

The air density r = 1.225 kg/m3. Based on Weibull distribution, the wind power density per unit is calculated as

$$P_W = \frac{1}{2}\rho c^3 \Gamma\left(1 + \frac{3}{k}\right) \tag{15}$$

4 Results and Discussion

4.1 Wind Velocity Frequency Distribution

The wind velocity (m/s) frequency distribution is depicted in Fig. 1 for all weather stations. The wind parameter frequency distribution helps answer the following questions: How long is the wind plant inactive due to insufficient wind speed? What is the scope of frequent wind speed? The wind velocity per hour is greater than 3 m/s between 30 to 40% of events at Ennore port, Karaikal, Chidambaram and Adhirama-pattinam weather stations. The hourly wind data is elevated to 40-50% of occurrences at Nungambakkam, Sathyabama, Anupuram, Thiruchendur and Neyyoor weather stations at the same speed. The calculated wind data on an hourly basis is above 3 m/s for 73% of the time at Thoothukudi weather station.

4.2 Wind Direction Frequency by Wind Rose Diagram

The energy prediction from wind source is fundamentally dependent on the reliability of the wind direction. In wind energy conversion study, wind direction is the most important parameter. During a wind turbine site selection, a depth study of wind parameter values at that location taken frequently is very important. The wind rose diagrams for weather stations shown in Fig. 2 indicate the variance in wind direction frequency distribution. From the wind rose diagram, the extreme wind directions were easily analyzed. The extreme wind directions in Ennore Port, Nungambakkam, Chennai and Sathyabama University were found to be between 210° and 270°.

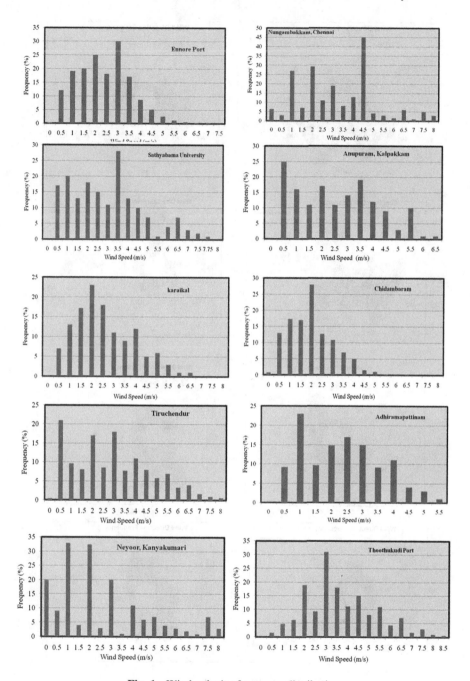

Fig. 1. Wind velocity frequency distribution

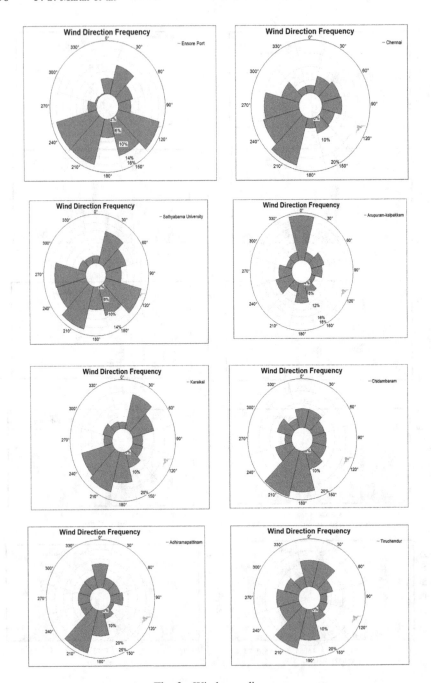

Fig. 2. Wind rose diagram

4.3   Weibull Parameter Analysis

Time series hourly Weibull shape parameter (k) and Weibull scale parameter (c) values are deliberated for all weather stations with the average wind velocity and wind energy density given in Table 2. Depending on the stipulation of the weather stations, the Weibull shape parameter (k) and Weibull scale parameter (c) values are considerably dependent on each other. It was noticed that the standard wind speed data from 2.2 m/s to a maximum of 3.8 m/s were very near to Weibull parameter values. For developing wind energy applications, the Weibull distribution model is extensively time-honoured to give good quality wind velocity data.

Table 2. Average wind energy density using Weibull parameters

Station	Weibull scale parameter (c) (m/s)	Weibull shape parameter (k)	Average wind speed (m/s)	Average energy density (W/m^2)
Ennore Port	2.9	2.7	2.8	19.6
Nungambakkam	3.5	3.1	2.5	17
Sathyabama	2.4	1.9	2.4	15.4
Anupuram	2.8	2.1	2.6	22.4
Karaikal	2.9	2.6	2.4	15.1
Chidambaram	3.8	2.3	2.2	10.5
Adhiramapattinam	4.2	3.2	2.3	12.7
Thiruchendur	3.0	2.0	3.1	39.4
Thoothukudi	4.0	2.6	3.8	57.3
Neyyoor	2.6	2.5	2.3	13.7

The following Fig. 3 shows the wind power density prediction using Weibull parameter analysis with Maximum likelihood and least squares method. These wind power values were compared with the actual data. It was proved that the Weibull analysis results are almost predicted with the actual data by the wind turbine generator.

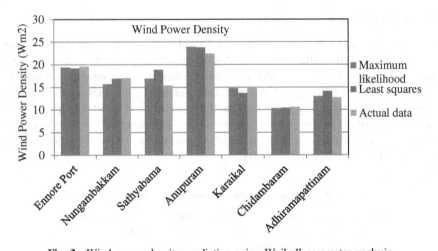

Fig. 3. Wind power density prediction using Weibull parameter analysis

5   Conclusions

Estimation of wind energy output for small-scale systems has been the subject of this paper. The main aim has been to use the Weibull-representative wind data instead of the measured data in time-series format for estimating the wind energy output. The Weibull function parameters were calculated analytically on a monthly basis, using the gamma function, from the measured data in time-series format. The wind speed data in frequency distribution format have been generated based on the Weibull distribution function, using the parameters identified earlier.

References

Zaharim, A., Najid, S.K., Razali, A.M., Sopian, K.: Analyzing the East Coast Malaysia wind speed data. Eur. J. Sci. Res. **32**(2), 208–215 (2009)

Celik, A.N.: A statistical analysis of wind power density based on the Weibull and Rayleigh models at the southern region of Turkey. Renew. Energy **29**, 593–604 (2003)

Jyosthna Devi, Ch., Syam Prasad Reddy, B., Vagdhan Kumar, K., Musala Reddy, B., Raja Nayak, N.: ANN approach for weather prediction using back propagation. Int. J. Eng. Trends Technol. **3**(1), 19–23 (2012)

Ghobadi, G.J.: Bahram Gholizadeh and Bagher Soltani: Statistical evaluation of wind speed and energy potential for the construction of a power plant in Baladeh, Nur, Northern Iran. Int. J. Phys. Sci. **6**(19), 4621–4628 (2011)

Gnana Sheela, K.: Computing models for wind speed prediction in renewable energy systems. IJCA Special Issue on Computational Science – New Dimensions & Perspectives, NCCSE, 3, 108–111 (2011)

Gokcek, M., Bayulken, A., Bekdemir, S.: Investigation of wind characteristics and wind energy potential in Kirklareli, Turkey. Renew. Energy **32**, 1739–1752 (2007)

Incecik, S., Erdogmus, F.: An investigation of the wind power potential on the western coast of Anatolia. Renew. Energy **6**(7), 863–865 (1995)

Carta, J.A., Velazquez, S., Matías, J.M.: Use of Bayesian networks classifiers for long-term mean wind turbine energy output estimation at a potential wind energy conversion site. Energy Convers. Manag. **52**, 1137–1149 (2011)

Choudhary, K., Upadhyay, K.G., Tripathi, M.M.: Estimation of Wind Power using Different Soft Computing Methods. Int. J. Electr. Syst. **1**(1), 1–7 (2011)

Gnana Sheela, K., Deepa, S.N.: Performance analysis of modeling framework for wind speed prediction in wind farms. Sci. Res. Essays **7**(48), 4138–4145 (2012)

Philippopoulos, K., Deligiorgi, D.: Application of artificial neural networks for the spatial estimation of wind speed in a coastal region with complex topography. Renew. Energy **38**, 75–82 (2012)

Giorgi, M.G.D., Ficarella, A., Tarantino, M.: Error analysis of short term wind power prediction models. Appl. Energy **88**, 1298–1311 (2011)

Naderian, M., Barati, H., Chegin, M.: A New approach for wind Speed Modeling. Bull. Environ. Pharmacol. Life Sci. **3**(1), 169–174 (2013)

Mohandes, M.A., Rehman, S., Halawani, T.O.: A neural networks approach for wind speed prediction. Renew. Energy **13**(3), 345–354 (1998)

Mohandes, M.A., Halawani, T.O., Rehman, S., Hussain, A.: Support vector machines for wind speed prediction. Renew. Energy **29**, 939–947 (2004)

Mortensen, N.G., Heathfield, D.N., Landberg, L., Rathmann, O., Troen, I., Petersen, E.L.: Getting started with WAsP 7. Riso National Laboratory, Roskilde (2001)

Benzaghta, M.A., Mohammed, T.A., Ghazali, A.H., Soom, M.A.: Prediction of evaporation in tropical climate using artificial neural network and climate based models. Sci. Res. Essays **7**(36), 3133–3148 (2012)

Brahmi, N., Sallem, S., Chaabene, M.: ANN based parameters estimation of Weibull: Application to wind energy potential assessment of Sfax, Tunisia, International Renewable Energy Congress, 5–7 November (2010)

Al Buhairi, M.H.: A statistical analysis of wind speed data and an assessment of wind energy potential in Taiz-Yemen. Ass. Univ. Bull. Environ. Res. **9**(2), 21–23 (2006)

Yilmaz, V., Celik, H.E.: A statistical approach to estimate the wind speed distribution: The case of Gelibolu region. Dogus Universitesi Dergisi **9**(1), 122–132 (2008)

Bechrakis, D.A., Sparis, P.D.: Correlation of wind speed between neighboring measuring stations. IEEE Trans. Energy Convers. **19**(2), 400–406 (2004)

Damousis, I.G., Alexiadis, M.C., Theocharis, J.B., Dokopoulos, P.S.: A fuzzy model for wind speed prediction and power generation in wind parks using spatial correlation. IEEE Trans. Energy Convers **19**(2), 352–361 (2004)

Oğuz, Y., Guney, I.: Adaptive neuro-fuzzy inference System to improve the power quality of variable-speed wind power generation system. Turk. J. Elec. Eng. Comp. Sci. **18**(4), 625–646 (2010)

Rasit, A.T.A.: An adaptive neuro-fuzzy inference system approach for prediction of power factor in wind turbines. J. Electr. Electr. Eng. **9**(1), 905–912 (2009)

Kumar, V., Joshi, R.R.: Fuzzy logic based light load efficiency improvement of matrix converter based wind generation system. Journal of Theoretical and Applied Information Technology **3**(2), 79–89 (2007)

Author Index

Printed in the United States
By Bookmasters